THE

TECH

WRITING

GAME

THE
TECH
WRITING
GAME

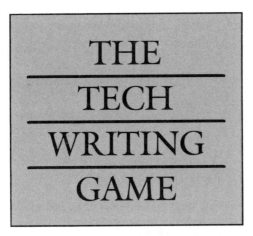

Janet Van Wicklen

Facts On File
New York • Oxford

The Tech Writing Game
Copyright © 1992 by Janet Van Wicklen

Facts On File, Inc.
460 Park Avenue South
New York NY 10016
USA

Facts On File Limited
c/o Roundhouse
Publishing Ltd.
Oxford OX2 7SF
United Kingdom

Library of Congress Cataloging-in-Publication Data
Van Wicklen, Janet.
The tech writing game / Janet Van Wicklen.
p. cm.
Includes bibliographical references and index.
ISBN 0–8160–2607–6 (acid-free paper)
1. Technical writing—Vocational guidance. I. Title.
T11.V37 1992
808.0666'023—dc20   91–41591

A British CIP catalogue record for this book is available from the British Library.

Facts On File books are available at special discounts when purchased in
bulk quantities for businesses, associations, institutions or sales
promotions. Please call our Special Sales Department in New York at
212/683-2244 (dial 800/322-8755 except in NY, AK or HI) or in Oxford
at 865/728399.

Text design by Donna Sinisgalli
Jacket design by Sam Moore
Composition and manufacturing by the Maple-Vail Book Manufacturing Group
Printed in the United States of America

10 9 8 7 6 5 4 3 2 1

This book is printed on acid-free paper.

■ ■

CONTENTS

APPENDIXES

ANNOTATED BIBLIOGRAPHY   226

INDEX   233

■ ■

CHECKLISTS

■ ■

PREFACE

When I first considered becoming a technical writer, I wanted to know not only what technical writing is but also how a tech writer really spends his or her time. I requested informational interviews with writers, visiting them in their work environment and listening to their candid opinions on the profession. From these interviews, I concluded that I would try technical writing.

I would like to give you the advantage I had of knowing what I was getting into. Consequently, the first part of this book takes you on an informational interview, with me and other writers, so you can acquire a "streetwise" view of the profession. From the impressions you gather, you can conclude whether or not it's for you.

If you decide to proceed with a tech writing career, the rest of this book guides you through the technical publication process and on through business skills you'll need to advance your career.

To read this book is to take a tour of the kinds of experiences a new technical writer can expect. Along the way, you will receive practical guidelines for developing the skills required by this dynamic profession. These skills include active listening and logical thinking, as well as writing.

This book differs from other books about technical writing in that it is not just about writing. Although it offers detailed guidance about writing and editing, this book is as much about communication, because interview and communication skills are central to technical writing.

This book is also about doing business, especially because technical writers are most often either creative people or technical people who by nature may not relish the politics of the business world. However, as a technical writer, you sometimes must play diplomat between marketing and engineering factions that enjoy less than congenial relationships. You also must understand the larger needs of a company, in order to best serve them. And at some point you may face the question of going into business as an independent consultant. Then you need to know about marketing your services, drawing up a contract, and setting your rates. This book addresses the business questions a serious technical writer will face.

Good writing and editing habits are the heart of the technical writer's craft. In this book, writing and editing are treated as the critical skills that they are, and are complemented by an emphasis on quality: proofreading every draft, proofreading format as well as grammar and spelling, and working with artists to create a visually appealing document.

In this guide, you'll find tips on how to use the latest technology available to technical writers: desktop publishing. This fascinating world of publishing capabilities puts added demands on the technical writer, requiring you to know about type fonts and other format elements that formerly were the concern only of typesetters, visor-capped proofreaders, and printers.

Lastly, this book discusses where to go from here. What career paths are available to the technical writer? You will find that many paths lead toward professional enrichment: careers in writing, management, technology, and more open up to the successful writer.

You will find my biases as a technical writer have shaped some of the material and terminology in this book. I am predominantly a software writer, specializing in data communications and icon-based user interfaces. What a mouthful! What that means is that I write about computer programs that allow different computers to communicate with each other, sharing information, and about computers that represent certain concepts in pictures, instead of words. What that means about this book is that I call technical people "engineers," product developers," and just plain "developers," instead of "physicists," "technicians," or any of a number of titles applied to the people who will be your primary sources of technical information.

I think that the principles of communication and clear writing are universal. I have sought to aim this book at as diverse a readership as possible.

I became a technical writer out of a desire to do creative work and still make good money. My objective as a writer, and my primary joy in my work, has always been to grow in the craft of writing. It is that objective which has led to the writing of this book.

■
ACKNOWLEDGMENTS

In writing this book, I've had to depend on so many people for help, advice, and information. I want to begin by thanking those writers, publications managers, and graphic artists who allowed me to interview them. Their real names appear herein, unless they requested otherwise. I believe their worldly advice enlivens these pages more than any other aspect of the book.

I could not have written this book without the help of those who have made superior contributions to the field of technical writing: Jonathan Price, R. John Brockmann, William Horton, Dr. Janice Reddish, Dr. Stephanie Rosenbaum, and the many others I have learned from in the course of writing this book.

I am also grateful to Dr. Frank R. Smith, Editor of *Technical Communication,* Journal of the Society for Technical Communication, for permitting me to quote freely from the journal's pages.

Thanks to those who reviewed the proposal for this book and whose suggestions helped shape it: Jonathan Price, Dr. Thomas Pearsall, John Huber, Meg Morris, and Dirk van Nouhuys.

Special thanks goes to Jonathan Price, who generously returned my phone calls and acted as my mentor through the inception and sale of the proposal for this book, and who has answered my most obscure questions about the mysterious world of commercial publishing.

Thanks to my agent, Carol Mann, who understood the book and quickly found it a home. At Facts On File, thanks to Susan Schwartz for her guidance and encouragement.

PART ONE

WHAT AM I GETTING MYSELF INTO?

■ ■

WHY BEGIN?

You picked up this book because you're curious about technical writing. Either you're a new tech writer or you're considering taking the plunge and becoming one. But first, you want to know what you've gotten—or are getting yourself—into. What are the rewards in this field? Will they satisfy you?

The book in your hand is a comprehensive guide to the field of technical writing. It gives you a chance to look before you plunge; then guides you through the day-to-day realities of being a tech writer. It includes tips about handling on-the-job situations not usually discussed in technical writing books, such as how to get information from a reticent engineer and how to deal with office politics in a high-tech environment.

You'll learn from working writers what they do from nine 'til five, what they find rewarding and frustrating, and how they broke into the field. You'll learn how to research and produce a technical document, including interviewing engineers, scheduling a writing project, and planning illustrations. You'll learn about professional associations and career development.

You can use this book both to decide about a tech writing career and to begin one. But first . . .

WHAT IS TECHNICAL WRITING?

Stated very simply, technical writing involves translating technical ideas into words a specific audience will understand. Technical ideas are initiated by scientists or engineers.

For example, a computer incorporates technical ideas that originated in an engineering laboratory. Engineers use combined knowledge of electronics, programming, and usability studies to come up with the optimal shape and internal arrangement for the computer.

Once the computer is produced, its operation must be communicated to the audiences that will market, repair, and write programs for the computer. At least three technical writers might be involved. They'll take notes on the computer's characteristics and create documents: One will write marketing brochures, another will write a hardware troubleshooting manual, and a third will write a software reference manual.

It is up to a technical writer to weave a connection of words between the scientist and his or her audience. And most often, this connection includes images. Technical writers not only write, but frequently design the whole document. They think up visual ideas to illustrate technical ideas. They provide tables, graphs, and simple drawings and work with artists to create more elaborate drawings.

The technical writer's audience almost always knows less about the subject than the scientist or engineer initiating the ideas. This means that the technical writer has the job of (1) understanding technical ideas initiated by a scientist, (2) understanding who the reader is, and (3) expressing the ideas according to that reader's education level, ability to understand, and familiarity with the subject.

Linda Lininger, a successful Silicon Valley free-lance writer, sees technical writing as a craft:

> I look at technical writing as being more of a craft than a profession; it's like being more of an artisan than an artist, because technical writers more or less make shoes. Either you make O.K. shoes or you can make really fine shoes. They both do the job, but some are better to look at, they feel better, they're pretty, and people go "Ooo Ah." The [recently finished manual]—I really put my heart and soul into it. Not only did I have a peak experience doing it, because it was a lot of work, pressure, drive, and crunch crunch, but it's nice to look at.

In the sense that Linda is talking about it, technical writing challenges the writer both to communicate effectively and to create a quality product. In her situation, a demanding schedule added further challenge. We see a picture emerging of technical writing as something of an adventure.

Technical writing is an adventure in yet another sense, in that technical writers are at the scene of many scientific breakthroughs. They enjoy the privileged knowledge of insiders. When technology allows humankind to take another giant step, a technical writer is there to learn and tell about it.

> Writers have supported and reported the nation's space programs, from its weather, communication, and navigation satellites to its Herculean effort to place man's footsteps on the moon . . . Writers have supported and reported the nation's atomic energy program, from the excitement of the first self-sustaining chain reaction to the development of nuclear power plants. In the future, writers will document and describe the nation's battle to control pollution, its struggle to save the cities, and its search for alternative transportation systems. They will continue to report man's fight against disease and disability. Where the action is, there the writer is.[1]

Thus wrote Clark Emerson and Vernon Roote in their 1972 career guide, *Your Future in Technical and Science Writing*. Since 1972, science has produced more wonders than Emerson and Roote could have anticipated, from heart transplant operations to laser weapons in space. And for each wonder, writers have participated.

But, what kind of people become technical writers, what do they really do from nine 'til five, and where did they learn how to do it?

WHAT'S IN IT FOR ME?

> It really is terrific being an insider, knowing about exciting new products that are coming, before your neighbors do. That's part of working at any high-tech company. I'm sure that people working in the design studios of automobile companies feel the same way or people in the movie or record business . . . I know about things that the whole industry is going to be talking about or the whole office is going to be talking about—later. I know about them now.

With these words, John Huber at Apple Computer describes one of the satisfactions of being a technical writer. Another reward he describes is "being published, without the uncertainties of being a free-lance writer." And for him, as for me, being paid well to learn is another big reward.

Every tech writer has a different list of payoffs they find in their job. Most writers include:

- producing a tangible result.
- being published.
- being paid well.
- meeting interesting people.
- learning new technologies and skills.
- enjoying writing.

TANGIBLE RESULTS

Technical publications manager Susan Tisdale reminisces about her writing days when, in contrast with her current job, her accomplishments were in print:

> You have a finished product at the end of your efforts. You have something that is the fruit of your labor, and it's all printed and nice, and it's going out with the product. You can hold it in your hands and say, "I did this."
>
> [When you're a manager] you solve problems, you talk to people, you write performance evaluations, so you never have a finished product that you can hold and say "I did this." Especially in retrospect, that's something that I always liked; that I never realized how much I liked until I left it.

Linda Lininger also appreciates this aspect of technical writing:

> It's not like some jobs where you work and work and work and have nothing to show for it. When you work, it's in black and white . . . There's something you can touch.

BEING PUBLISHED

Being published is a big tangible result. A writer whose book is published before setting pen to paper is rare. In the commercial publishing world, a writer goes through queries, proposals, and negotiations, before even the prospect of a monetary advance comes into view. To be automatically published is a privilege enjoyed—nay, taken for granted—by technical writers. Although a technical writer's name usually does not appear in the book, and some writers have strong feelings about this omission, a published book is still a source of pride:

> When this manual comes back from the print shop—my very first manual all by myself—when that comes back from the print shop, that'll be a hell of a wonderful day. It'll be time to buy a bottle of champagne. I'm very object

oriented. I want to produce some*thing*. For this manual, I'm handling every-thing—the writing, vending it out to an editor, getting the art done. I'll be handling it through the print shop. So I'm the writer and the production editor, too. I think that's a hell of a production.

These comments point out how a technical writer often needs to be a Jack-of-all-trades, and this writer obviously enjoys that aspect of her job.

GOOD PAY

Getting paid well to write, while you are writing, is another privilege of the profession. A 1988 survey of technical writers' salaries, conducted by the Society for Technical Communication (STC), revealed an annual median income of $34,000 for that year. An earlier STC survey (1985) showed salaries varied in relation to geography, industry, and, all too predictably, sex. New York and Pennsylvania paid the highest; Indiana, Kentucky, Michigan, and Ohio the lowest. Government and military programs paid more than private industry. The 1988 survey shows salaries in private industry caught up with government. However, both surveys revealed men got paid more than women.

Technical writers' salaries grew at a faster rate in the early '80s than since 1985. Nonetheless, if they continue to grow even at the slower recent rate, a tech writer working in 1995 can expect to make about $44,000 a year and a senior writer significantly more. Some Silicon Valley senior writers in 1991 were earning over $50,000.

It's not unusual for a technical writer in the computer industry to double his or her salary in the first year. Most tech writers I interviewed did so. And if a book gets canceled, for whatever reason, the staff technical writer still gets a regular paycheck.

The *1991 Writer's Market* reported free-lance technical writers' hourly pay between $35 and $75, "depending on degree of complexity and type of audience."[2]

INTERESTING PEOPLE

Meeting different kinds of interesting people is another plus. How many professions put you in contact with successful engineers, artists, marketing wizards, and of course, other writers? These contacts can provide product information, as well as friendships and possible references when it's time for a job change.

When John Huber wrote a book called *The Human Interface Guidelines,* he needed to explain the aspects of computer design that make computers easy for people to use. To do research for this book, John spent a whole year working with a group of psychologists and graphic artists. The group traveled together, used the research facilities of a number of major universities, and enjoyed an enriching synergy.

LEARNING NEW SKILLS

Another reward in technical writing is the opportunity to constantly learn about new technologies and better ways to communicate. Says Billie Levy, a writer who specializes in documenting computer hardware:

> One quality that a technical writer ought to have is a joy in learning. I spent a lot of time going to college, and I didn't care what college class you dropped me into, as long as it had a good teacher and some good material. It could be on some subject that it never occurred to me before to be interested in. But once I'd start working on it, I'd get interested. It's part of my personality that I love to learn. And that's one of the reasons I got into this business and that's one of the reasons I love learning about this machine. I love learning about new things.
>
> If you want to do the same old thing day after day, don't be a writer. Some people like to be on automatic pilot. They like to learn how to do a job, and learn how to do it so well that they don't really have to put a lot of energy into learning new things. They just keep cranking. That's not me and that's not technical writing.

Besides learning new technologies, tech writers get to learn the latest word-processing tools. And they're repeatedly challenged to learn better communication skills, because they need information from people and they need to harmonize as part of a team that produces a product.

ENJOYING WRITING

One reward in being a professional writer is the enjoyment of writing. This seems obvious, but it's a subtle truth. Most of what technical writers do is not writing, but research, organization, learning, human interaction, and book production. I feel the writing time is really the appetizer or dessert course of a meal composed of grittier dishes. It's my favorite part of the meal, which is why I'm writing this book!

Being published, receiving good pay, working with interesting people, learning new concepts and skills, and enjoying writing are probably the major benefits writers find in their jobs, but most writers find others as well. If these are things you care about, or could come to care about, you'll experience tech writing as a richly rewarding career.

However, there's a catch. Technical writing is considered by many to be one of the most stressful professions today.

 ## WHAT PROBLEMS COME WITH THE TERRITORY?

"On the one hand, you're told to be complete, correct, concise, perfect, personal, accurate, and on the other hand, finish it by tomorrow."

Thus one tech writer sums up what he considers the stresses of the job. Every tech writer has their own grief list. Many will include:

- difficulty obtaining information.
- reticent or uncooperative engineers.
- canceled projects (after the work's been done).
- unreasonable or unclear deadlines.
- unwieldy tools and equipment.
- office politics.

DIFFICULTY OBTAINING INFORMATION

Difficulty getting product information is a common experience among writers who've been at it a while. Sometimes the product is not finished yet, and its developers have not had time to write down anything about it. Sometimes a product developer is the only source of information and is difficult to communicate with.

According to one technical writing manager, who prefers anonymity:

> There is a developer here who is quite a character. He doesn't focus on the technical information but rather uses the interviews with the writer to express his political opinions about the company. So he gets up on his high horse, "At [this company], we'll put any product out there for anybody, and we never have a chance to do any of them right." The frustrating thing for the writer is trying to get technical information and keeping this guy focused on what he's supposed to be telling them, rather than expressing all this vehemence against the way the company's run.

This manager sympathizes with her writers' difficulties communicating with this engineer and understands when they need extra time to get the information they need from him.

RETICENT ENGINEERS

I remember meeting with a man I'll call Max Engineer who was my primary source of information about a computer networking product. I asked him a series of well prepared, detailed questions about the product, and he answered all my questions with "probably" and "I'm not sure." No matter how I rephrased my questions, Max was not forthcoming. Later my boss met Max at an office party and happened to mention to him that he was my key source of technical information. Max was shocked. He thought writers pull information from air, like magicians pulling rabbits from hats.

In that particular situation, I was too new to technical writing to know that I could have educated Max. I could have told him a bit about what technical writers do and informed him of his crucial role in that process.

According to computer-hardware writer Billie Levy:

> There's no point in browbeating the individual engineers. You have to go upstairs and say, look, if you want this manual done, it has to be looked at. It has to be passed by those people whose O.K. is important.
>
> Problems getting information are endemic to our profession. The engineers are very busy and they have a lot of demands on their time. So you have to be persistent and tactful.

The chapter "Communicating with Engineers" describes some effective ways to deal with reticent, uncooperative, and just plain overworked engineers.

CANCELED PROJECTS

I count canceled or postponed projects high on my list of the nonrewards of technical writing. After putting months of work into a manuscript, I anticipate its completion somewhat as a pregnant mother anticipates delivering a baby. When a project miscarries, for whatever reason, the next project seems harder to begin.

Technical writing manager Susan Tisdale sees postponed projects as a result of shifting priorities:

> In a company like ours, you have so many projects and so many products going on at the same time, and you have a limited number of writers. So you have writers working on things based on an understanding of what the priorities are. The most important products get their attention first.

Susan believes it's counter-productive "for a writer to get ramped up on a product, become productive writing, get to know the other team members, the technical people, and all that; then be yanked off of that onto whatever is next week's fire—'Oh well, we have to do some preliminary documentation because they want to send something out to a customer, so drop what you're doing and go . . . throw together something.' "

Because Susan sees her job as "putting people in a position where they can succeed," she gets as frustrated as the writer when projects get "reprioritized."

WE WANT IT GOOD AND NOW

The worst thing about technical writing that happens again and again is this business of having to meet a deadline when you sense that the project isn't going to be ready and your book doesn't need to be ready. The deadline is unreal, the boss is holding you to it only to make himself look good, or the deadline slips and it just starts all over again. That's the worst part.

With these words, a computer-software writer describes his real feelings about unreal deadlines.

Nowhere else is the deadline so malleable one day and unforgiving the next as it is in fiercely competitive computer companies. The book must be ready *now;* then the product is delayed and anyway, two reviewers think the book needs a rewrite.

Unfortunately, deadlines are a high-tech fact of life. The push to get out the latest computer, car, VCR, or biomedical breakthrough is the frenetic reality of our competitive marketplace. It's why we have wonder toys and miracle cures. If you're going to write about them, you sometimes have to write on your feet, as you follow the product out the door.

TOOLS WON'T WORK

Most tech writers have experienced the printer breaking down in the heat of a deadline. Office jokes abound about machines that sense panic and refuse to cooperate. Machines do tend to break down when they're used more heavily than usual, which happens around deadlines. So there's good reason for such "crashes." They are probably not the act of a rebellious machine consciousness or of demons residing within. (I say "probably.")

Other frustrations with tools occur because of poor maintenance or bad choices. The computerized workplace is like an ecosystem. It contains an intricate balance of software and hardware that must interact. This environment requires a knowledgeable system administrator to keep the machines talking to each other and listening to the people who use them. If a computer uses the wrong "protocol" (set of rules) to communicate with a new printer, the printer won't work. If the people who use machines don't choose and maintain them carefully, production suffers.

Technical writers are at the mercy of these office machines and are also vulnerable to the failings of their computer software tools: text editing and formatting software. Some older software requires the writer to type detailed computer commands within a manuscript. The formatting program then reads these commands and prepares the manuscript to be printed. Modern desktop publishing is rendering this two-step process obsolete by allowing the writer to see on the computer screen what the final printout will look like. But many archaic tools are still in use, with the added disadvantage that fewer and fewer people know how to use them. Often these tools required preliminary programming to work properly, and the programmer who did this is no longer around to teach current staff the idiosyncrasies of the system.

Technical writing requires patience with tools. If you are comfortable with the personality of computers, all the better, because you will spend some time "debugging" your tools.

THE WRITER DIPLOMAT

Office politics provide another source of stress. For example, a technical writer can become caught in the middle of a miscommunication between departments. Marketing says the product is one way, engineering knows it's not, and neither is speaking to the other about it. The tech writer must play diplomat, clarifying the communication between these two departments.

Occasionally, powerful personalities can make a writer's job difficult, if not impossible. At one glamorous company known for its wonderful documentation, a writer acquaintance (who requests his name not be used) was assigned to write a book about a data communications product. He did not know when they gave him the assignment that a powerful engineer did not want the book written. The engineer had already written one about the product—a book which was almost unreadable. He liked his book and wanted to see it published with his name.

The writer tried to create a reasonable schedule and write a high-quality manuscript, but his efforts were thwarted. His manager kept encouraging him to leave the engineer's prose intact. My acquaintance eventually detected the pressure being exerted by the engineer, chose not to take the matter personally, and requested a different project.

Another way to handle this pressure would have been to leave the engineer's prose alone. However, if you are idealistic and care about your work, such compromises can lead to burnout (discussed in the chapter "The Hazards of Being a Tech Writer," later in this book).

To withstand the criticisms, pressures, and obstacles of technical writing requires emotional distance from the day-to-day crises. And, it requires a sense of humor. Fortunately, most writers have a sense of humor, and their camaraderie is one of the joys of the profession.

A MENTOR CAN HELP

You will have more success dealing with stress if you have a mentor—a business acquaintance, perhaps even a boss, who has been there before. A mentor can tell you when a crisis is just typical silliness resulting from a bad combination of business pressures and human nature. "We have no control over the lunacy around us. Remember that," my boss remarked just today.

If you are not fortunate enough to have a mentor and you haven't yet developed suitable caution, you might fall prey to some of these pressures. You might suggest a more realistic deadline to a manager too harried to listen. You might suggest an important improvement for a rushed book that no one really cares about. These suggestions can make you unpopular. Take heart! Almost every seasoned tech writer has had these experiences. As one colleague put it, "In this business, you're either a bum or a hero."

Every writing job has limits and opportunities. If polished writing isn't possible in a particular writing department, because of deadlines, politics, or uncooperative developers, you can find other ways to grow and learn. Maybe you can acquire technical knowledge by attending in-house training courses. Perhaps the company will sponsor you as a part-time student at a community college, where you can acquire technical skills that will help you understand and write about the company's products.

In one particularly frustrating job, where equipment never worked and engineers were always too busy to explain things, I took company training classes in everything from beginning data processing to "packet switching," a complex technology that prescribes how computers transmit data.

The tech writing field is far more an opportunity to expand yourself and grow than it is a series of frustrations, although sometimes it may seem the latter. If you distance yourself and learn to recognize the silliness of typical crises, you'll be the calm one others turn to for perspective.

The chapter "The Hazards of Being a Tech Writer," later in this book, goes into greater detail about ways to deal with the stresses of a technical writing career.

SUMMING IT UP

This chapter talked about some of the rewards of tech writing and some of the grief. Now you know more about what you're getting yourself into. If you're still game, read on. In the next chapter, you'll learn about the history of technical writing and its emergence as a profession.

▪ ▪

[1] Clark Emerson and Vernon Roote, *Your Future in Technical and Science Writing* (New York: Richard Rosen Press, Inc., 1972), p. 21.

[2] Glenda Tennant Neff, ed. *1991 Writer's Market* (Cincinnati: Writer's Digest Books, 1990), p. 36.

2

A LITTLE HISTORY

W e have now made clear how the axle is to be constructed;" wrote Hero of Alexandria in the first century A.D., "so now we shall describe its use."

Having explained how to make a winch, Hero added these user instructions, which are probably the first in human history: "If you want to move a great burden by a smaller power, you fasten the ropes that are tied to the burden on the grooved places of the axle on both sides of the wheel. Then you put handspakes into the holes that we have made in the wheel and press down on the handspakes, so that the wheel is turned, and then the burden will be moved by a smaller power . . ."[1]

Hero's instructions exemplify good user documentation, at least in this translation from A. G. Drachmann's *The Mechanical Technology of Greek and Roman Antiquity*. They address the user directly; they're simply phrased and clearly organized.

Through most of history, scientists have documented their work for fellow scientists. However, examples of technical writing for consumers exist from as early as 1748 in America, when the English handbook, *The Instructor*, was revised for American use. The handbook is described in Evald Rink's *Technical Americana* as "Containing spelling, reading . . . Together with the carpenter's plain and exact rule: shewing how to measure carpenters, joyners, sawyers, bricklayers, plaisterers, plumbers, masons, glaziers and painters work. How to undertake each work, and at

14

what price . . . Likewise the practical gauger made easy: the art of dial-
ling . . . instructions for dying, colouring, and making colours . . . The
whole better adapted to these American colonies . . ."[2] This manual's
reader was the original jack-of-all-trades! Were its author alive today, he
wouldn't believe how far and fast technology has advanced—nor how
specialized it's become

In the 1700s, technical information about industrial processes and
machinery was called "useful knowledge." Such knowledge was imparted
in lectures to early professional associations, like the Providence Association of Mechanics and Manufacturers, which formed in the late 1700s.
But technical writing was not recognized as such until well into the 20th
century.

The earliest book specifically on technical writing is probably Sir
T. Clifford Allbutt's *Notes on the Composition of Scientific Papers,* which he
wrote for his students at the Faculty of Medicine in Cambridge in 1904.
The writing in Allbutt's profession appalled him, and he wanted to correct its defects in his students' work. "The prevailing defect of their composition" wrote Allbutt in his preface to the third edition (1923), "is not
mere inelegance; were it so, it were unworthy of educated men; it is such
as to perplex, and even to travesty or to hide the author's meaning."[3]
The good doctor would find similar flaws in the writings of scientists
today.

In the floweriest Victorian English, Allbutt offered advice about plain
language: "Let him not search afield for long and complicated forms and
elaborated words, nor for large and decorated vestures; if he can get well
home on his ideas the simplest and closest words are best. Let him consider not how finely, but how plainly and directly he can express himself." He then goes on for several long sentences before concluding, "If
the essayist, stripping off all encumbrance, will look nearer home for his
words, and put these together concisely, the figure of his thought will
move more freely in the lighter garment."[4] By modern standards, he was
a long-winded fellow, but his heart was in the right place.

Almost all early books about technical writing were written for the
education of engineers and scientists, according to Gerald J. Alred et al.
in *Business and Technical Writing: An Annotated Bibliography of Books, 1880–
1980.* "As far as we can determine," say the authors, "the first technical
writing book aimed strictly at a professional audience is Thomas Arthur
Rickard's *A Guide to Technical Writing* (1908). This work, aimed at met-
allurgical engineers and geologists, has a modern flavor throughout, especially with its emphasis on audience."[5] By professional audience, the
authors mean professional scientists, not writers.

THE TOWER OF BABEL

Until the 1940s, there were no technical writers. Technical products were primarily designed for technical experts and technical experts wrote any instructions or explanations that went with these products. So scientists and engineers wrote for each other, or more precisely, for themselves. They took pride in their use of scientific jargon and sometimes affected a high tone. Long words, sentences, and paragraphs enhanced the scholarly effect (or affectation, if you will). Often technical documentation produced by one scientist proved incomprehensible to another.

The fact that jargon and complexity fail to communicate was lost on scientists and engineers trained within a tradition of obscure verbiage. Robert Gunning, in *The Technique of Clear Writing,* tells the story of a great American scientist, Willard Gibbs, many of whose discoveries remained unknown because his colleagues couldn't understand his writing. Says Gunning, "It became a scientific joke that it was easier to 'rediscover Gibbs than it was to read him.' " Gunning quotes Gibbs:

> For the equilibrium of any isolated system it is necessary and sufficient that in all possible variations of the state of the system which do not alter its energy, the variations of its entropy shall either vanish or be negative.[6]

What did the man say? This is a good example of poor technical writing. It contains an unclear subject, unclear referents, and passive voice. The long sentence wanders into obscurity, as did its author.

During World War II, science and technology increased in importance and so did scientists. Because of their need to work unhampered by the demand for documentation, scientists acted as unwitting midwives in the birth of a new profession. They began requiring writers to interpret their creations to others.

"When I was in the Second World War," remembers 65-year-old tech writer Mark Smith, "I was working on aircraft using military specifications as sources of information, and sometimes the information was quite clear. But sometimes it wasn't, and I didn't know what I was reading. The choice of words and the way it was put together were clumsy, and perhaps a picture would have done better. But there wasn't one." Although Mark was not a writer at the time, he remembers thinking that if he ever got the chance, he would write better "specs" than the ones he had to work from.

World War II led to a huge increase in the number of military contracts. After the war, the government required that these contracts be accompanied by documentation. Additionally, the peacetime uses for war-related technology fed a technological boom in consumer products, which also required documentation.

"Usually the engineer kept an engineering notebook on his work," says Mark Smith. "It was the writer's job to come along and pick up on the facts to tell what the product was for and how to use it. And that developed into what's called an operator's manual."

Operator's manuals were written by the new breed of people who were hired just to write. "During the war and shortly after the war," says Mark, "it started more or less as just a documentation effort and then developed into technical writing. I know at the time, I never heard of a technical writer. I heard of writers, but not technical writers."

 ## THE BIRTH OF A PROFESSION

As technical writers increased in number, they sought validation for their profession by forming associations. Between 1953 and 1957, writers formed half a dozen professional societies around the country, including the Society of Technical Writers (1954) in the eastern United States, the Association of Technical Writers and Editors (1953) in New York City, and the Technical Publishing Society on the West Coast. The Society of Technical Writers and Editors (STWE) was formed out of the two eastern societies in 1957.

In the late '50s and early '60s, the Vietnam War again placed a demand on military industries, the number of military contracts increased, and more technical writers were required.

The late '60s saw the first graduating college class of technical writing majors, which at the time was quite an oddity. Scientists were perplexed by this group of graduating engineers who had made a commitment to the writing profession. And engineers they were, heavily trained in science and technology. They entered the business world with many of the biases of earlier scientists—that technical writing should sound formal and be written in third person and passive voice.

Fortunately for everyone, technical writers today are much more aware of the requirements of clear communication than they were in the '60s and '70s. What has brought about this change? The commercialization of technology.

THE COMPUTER REVOLUTION AND BEYOND

The biggest commercialization happened when the computer became widely available for business use in the '70s. The second most revolutionary development happened when it became widely available to consumers in the '80s—industry needed to sell its genius to us, the public.

When the microcomputer (the forerunner of personal computers) first appeared, it was a very mysterious thing indeed. If your computer could not successfully process information, it displayed a number, which corresponded to an error message. You could look the message up in a book, but you'd be lucky to understand what you found. Then computer programmers started writing messages in English—messages like "SOURCE IS READER" and warnings like "BDOS ERR ON A: SELECT." These messages were understandable only to programmers.

Office automation and the commercial availability of personal computers put pressure on industry to produce larger quantities of documentation. And the audience had changed. Technical people, and those selling their creations, came to realize they needed to explain technology to the nontechnical public—the "end user"—and they needed someone other than a computer programmer to make their documents understandable.

Enter the professional technical writer—someone who knows how to use language and graphics to teach technical concepts to nontechnical people. When the computer industry began to expand, or rather explode, the demand for professional technical writers also exploded.

Today, technical writing has evolved into technical communication. Writers in all fields now have access to sight-and-sound media. Sometimes writers design computer tutorials for end users who will never pick up a document. Their computer teaches them, through graphics and sound, all they need to know about how to use it. Computer technology today allows you to record, store, and send voice messages inside documents; to compose music; to manipulate full-color, three-dimensional objects in space; and more. Technical writers both describe these products and use them to communicate technical information.

So you can see the technical communicator's role is changing as technology changes, and has already evolved well beyond the role of writer. The field continues to evolve and grow rapidly.

According to the Society for Technical Communication's *1991–1995 Strategic Plan,* "the number of technical communicators in the United States is estimated to be more than 100,000 and will grow to 150,000 by 1995." With the proliferation of audio and video technologies, discoveries in biotechnology, new applications based on microprocessors, and other technological products, an unimaginable number of new opportunities will become available to technical writers in the next decade.

 ## SUMMING IT UP

In this chapter, you delved into the history of tech writing and found that it is very old. You learned about the relatively recent birth of technical writing as a profession and its flight from the printed page into realms of color and sound. The next chapter describes people who do well in this field and the kinds of skills required.

▪ ▪

[1] A. G. Drachmann, *The Mechanical Technology of Greek and Roman Antiquity: A Study of the Literary Sources* (Madison: University of Wisconsin Press, 1963), p. 56.

[2] Evald Rink, *Technical Americana: A Checklist of Technical Publications Printed Before 1831* (Millwood, N.Y.: Kraus International Publications, 1981), p. 18.

[3] Sir T. Clifford Allbutt, *Notes on the Composition of Scientific Papers* (London: Macmillan and Co., Limited, 1923), p. v.

[4] Allbutt, p. 140.

[5] Gerald J. Alred et al. *Business and Technical Writing: An Annotated Bibliography of Books, 1880–1980* (Metuchen, N.J.: Scarecrow Press, Inc., 1981), p. 2.

[6] Robert Gunning, *The Technique of Clear Writing* (New York: McGraw-Hill Book Company, 1968), pp. 254–5.

WHO BECOMES A TECH WRITER?

Who are successful technical writers? What are their personal qualities and work-related skills? The reason you might ask these questions, of course, is to find out if you are such a person.

Tech writers come from as varied educational backgrounds as you can imagine, from philosophy to computer science. The only background they seem to have in common is a degree of some sort, and even here, it's tough to generalize. A 1988 study, "Profile 88," conducted by the Society for Technical Communication (STC) surveyed the educational backgrounds of 600 randomly selected STC members. The results revealed that 88% of its membership had a bachelor's degree or higher.[1] This means that 12% did not.

I recently met a successful technical writer who has no bachelor's degree. While she's an exception, she also proves that if you don't have a degree and you think this field is for you, you can still go for it. (Hiring managers I interviewed agree, but stipulate that a job applicant with no degree needs lots more experience in the field than an applicant with a degree.)

The profile of a tech writer is changing. Hiring managers are beginning to perceive technical communication degrees as more desirable now that such degrees are widely available. Nonetheless, the high demand for technical writers, coupled with their history of diversity, makes for fairly permissive hiring practices. By permissive, I mean that many different kinds of backgrounds are acceptable to hiring managers. More about that in the chapter "Breaking In." This chapter analyzes the backgrounds, personalities, and skills of working writers today and gives you some ways to explore your own aptitudes.

 ## TECH WRITERS HAVE VARIED BACKGROUNDS

If you're an undergraduate deciding on a major, you might find it useful to turn to the next chapter, "Get Ready," and read the section "If You're Considering a Degree," before proceeding with this chapter. That section will give you a better idea of the degrees hiring managers now look for in new grads. If you already have a degree or are a career changer, read on.

Many—probably most—working technical writers are career changers who started in some other profession. Their educational backgrounds vary widely. According to the 1988 STC survey, most working technical writers are former humanities majors. Thirty-two percent come from English curriculums, while technical communication and journalism provide another 23%.

Jan Roechel, now a staff writer at Apple Computer, took a theology major and a journalism minor in college, and eventually got a job in the accounting department of a computer networking firm.

"When I was working in accounting as a supervisor, I worked with the programmers and I set up a computer system . . . I had an interest in programming, and I wanted to take courses so I could write programming applications for accounting." Jan's boss wanted her to go into accounting management and wasn't open to training her as a programmer. In scouting around the company for an opportunity to learn programming, Jan stumbled upon a technical writing job, realized it combined her writing skills and her interest in computer software, and took the opportunity to begin a new career.

Many technical writers are former teachers and feel that teaching contributed to their skills as technical communicators. "What the technical writer is basically doing is an educational process," says hardware writer Billie Levy. "A technical writer should be literate, and secondly, have a background in education. Because as a technical writer, you're essentially an educator. And education can teach you how to organize material to present it to other people so that they can absorb it."

When I asked publications manager Susan Tisdale what background prepared her to be a technical writer, she also stressed her teaching background:

> I think the skills are very similar. You're doing the same thing whether you're writing a manual or you're standing up in front of a classroom—You're collecting and organizing a body of information to present to other people so they can use it. There are a lot of parallels between the two. I think the skills I learned in my teaching probably helped me more than whatever writing training I've had.

The 1988 STC survey showed 7% of STC members have education degrees.

The physical sciences are well represented in the tech writing profession. The 1988 STC survey reported that 12% of its members majored in computer science, physics, chemistry, or math, and that another 6% have engineering degrees.

Programmers and engineers sometimes switch to technical writing after some years of industry experience. Sixty-five-year-old Mark Smith has been a technical writer for 36 years and an independent consultant for the last 11. He started as a service engineer in the aerospace industry.

> I was a service engineer for an F-102 and F-106 and also for the Atlas missile for about nine years. In that time, I became a service specialist in the hydraulic and pneumatic fields and I had to write the tech orders for the government on those systems. At the time I was doing that, I was also writing accident reports for an air force magazine and every time I'd write one I'd get an award—about $150.00. That was kind of an incentive. The feedback was that these were good articles, and so maybe I'm some kind of a writer [laughs] after all and don't know it! So I just kind of stayed with it, always writing engineering specifications and I seemed to have a knack for it.

About becoming a technical writer, a former quality assurance tester and customer support engineer says:

> I didn't see it as leaving the scientific training behind. As a matter of fact, the longer I stay in this business, the more I see that scientific training is very valuable for a tech writer. There's a lot of research involved. Training in scientific method and logical presentation of data or evidence are things that people who are exclusively oriented in the humanities sometimes are neither interested in nor capable of appreciating. I think a tech writer has to have a certain balance in their mental makeup and their approach to their job. Tech

writing, in one sense, is a literary activity, but I see it as a sort of scientific— or semi-scientific—endeavor.

These writers' stories give you some sense of the variety and permissiveness of the technical writing field. The chapter "Breaking In" describes how managers view requirements for this field and guides you on how to break in.

 ## WINNING QUALITIES

Despite their varied backgrounds, successful tech writers all have certain personal qualities and work-related skills. You may already possess many of these. Top of the skills list is the ability to communicate, not just on paper, but interpersonally and graphically. For this reason, many authorities, including the STC and the American Medical Writers Association, prefer the term *communicator* over writer. While I'll continue to use the terms interchangeably, I mention this to stress the interpersonal and visual communication skills required.

Now, here are those skills—successful technical writers are able to

- write.
- communicate interpersonally.
- communicate visually.
- tenaciously track information.
- learn quickly.
- take criticism.
- be flexible.
- meet deadlines.

WRITE

Technical writers can analyze, organize, and verbally communicate information so that it is easily understood by the intended audience. This requires both analytical skills and empathy for your reader. Says one senior writer:

> You have to be able to put yourself in somebody else's position. Especially if you're writing to end users—people who've never used a computer before—and you want to give them confidence. You have to sit in their chair and remember what it was like, and not assume too much knowledge.

COMMUNICATE INTERPERSONALLY

Technical writers need to communicate well to get information from engineers and other experts. And, as mentioned in the chapter "Why Begin?", writers are often political go-betweens. They need to be both diplomatic and assertive to get their work done. According to a writer at Apple Computer:

> A project team can be full of contradictory goals, and its members don't become aware of it until the book has to be written. And the writer is the one that has to say "Look, it's either that way or it's that way. You've gotta choose. It can't be both 'cause I have to write it down." And sometimes that's an unpleasant truth, and they don't like the writer for pointing that out to them. But you get to do it anyway.

COMMUNICATE VISUALLY

Technical writers create graphs and charts to illustrate technical concepts and they provide artists with ideas for illustrations. Additionally, writers usually must know how to use *desktop publishing* software (those computer programs that allow you to produce a typeset-looking document) and be skilled at manipulating the visual elements of a page design.

As mentioned in the chapter "A Little History," some technical communicators now design video presentations and graphically oriented computer programs, as well as presentations in other visual media.

Says biomedical writer Dan Liberthson:

> I spend a lot of my time doing slide sets to educate physicians and sometimes lay people in medical matters, and a slide has to be visually interesting. You do write an annotation—a legend—that goes with the slide, but the slide itself has to be visually gripping and interesting. You get some help from graphics people in the refinements, but the concept you have to come up with.

TENACIOUSLY TRACK INFORMATION

Technical writers are thorough researchers who can recognize holes in the information they've received. They're willing to pursue questions until they have answers, through any means available. Says one publications manager:

> You have to be a little bit aggressive, because there are going to be people who don't want to talk with you, who don't want to spend time with you,

and they don't want to review your manuals, either. So you can't be afraid to go and pound people on the head and say "I need this [information] or the manual is not going to be ready on time" or to ask your boss to do the same thing.

LEARN QUICKLY

Technical writers need to understand complex technologies enough to describe a high-tech product. Says Dan Liberthson:

> To be successful, I think you have to be very intensely project-focused. Learn like a sponge. You soak up like a blotter as quickly as possible, like cramming for an exam, everything that you need to know for a particular project. Then you do the project and then you let it go.

TAKE CRITICISM

Technical writers have their work reviewed to ensure its technical accuracy. Reviewers will say anything—even obscene things—about a writer's work. The chapter "The Review Process" goes into how to handle this in detail.

BE FLEXIBLE

Technical writers are faced with undefined product development schedules, canceled projects, and myriad other situations in which they are forced to rapidly switch gears. Says Tandem publications manager Carre Mirzadeh:

> Often at the beginning of a project, the project isn't well specified, the schedule isn't well known, the players on the project aren't well known, and you have to have a great deal of flexibility to be able to adjust to the requirements and constraints during the time of the project. And more than one writer I've known has been driven from technical writing because they weren't able to function in that type of an environment.

MEET DEADLINES

Technical writers need to manage an entire writing project. The final deadline, while affected by many forces outside the writer's control, must ultimately be in the writer's control. If a writer's schedule is jeopardized, he or she communicates that fact and renegotiates. Meeting deadlines requires every skill a writer has. Usually junior writers get help with this part of their job, and companies differ a great deal over how seriously they take deadlines. If you want to excel as a technical writer, meet your deadlines.

WHO ARE YOU?

Assessing your particular combination of personal qualities and work-related skills lies in the domain of vocational counselors. Some try to quantify "aptitude"—your talent in a subject area. Others believe it's your enjoyment and desire that best determine your success at something. In the latter camp, the classic is Richard Bolles' buoyant, extremely helpful *What Color Is Your Parachute?*[2] If you still have doubts about where you are going in life (who doesn't), read Bolles' book.

A number of career guidebooks present skills-assessment methods based on what you enjoy. They suggest listing significant achievements from your past that have afforded you the most personal satisfaction, then listing the actual work you did, the skills you used (such as organizing, advising, selling, etc.), and the rewards you experienced.

If you like more concrete skills measurements, see a vocational counselor and take an aptitude test. Or go through the exercises in Barry and Linda Gale's *Discover Your High Tech Talents.*[3] The danger in such tests lies in the conclusions you might draw from them. Don't conclude on the basis of a test that you cannot do something. Instead, interpret the results as an indication of where you may need to learn more. A low score on a numerical test might simply mean that you haven't dealt with mathematical concepts for years. You can correct this with practice.

Once you have a pretty good idea of your interests and abilities, use the "Winning Qualities" described in this chapter and the questions in the following checklist to decide if you are—or if you want to become—the right person for a technical writing career.

▪ ▪

Checklist 3–1.
Self-Exploration

The "correct" answers to these questions are pretty obvious. They're just here to help you think it over.

- Do you like to write?
- Do you like learning new things?
- Can you converse with "difficult" people and avoid getting into an argument with them?

- Have you ever negotiated a compromise between people who disagree?
- Have you ever drawn pictures, graphs, or diagrams to express concepts?
- When you can't get the answer to a question after several tries, do you think creatively about other avenues to explore?
- Do you try to either appease or argue with someone who criticizes you? Or can you step back and weigh their criticism objectively?
- If plans change unexpectedly, do you freak out? Or can you take a deep breath and steer accordingly?
- Do you keep your time commitments?

IT'S UP TO YOU

You can see that successful technical writers are self-directed, highly motivated learners, unintimidated by new concepts and terminology. You should have some of these qualities, if only in their embryonic form, before you consider tech writing. Many tech writers don't have these qualities, but they're not good tech writers.

This book assumes that you can develop any quality you want, if you really believe it's in your best interest. And none of these generalizations about tech writers is meant to intimidate. You can always find people to tell you how hard something is. This book is about how to go out and do it.

Says Richard Bolles in an early (1978) edition of *What Color Is Your Parachute?*, "Your interests, wishes, and happiness determine what you actually do well more than your intelligence, aptitudes, or skills do."[4]

If you don't have a skill you need, you can learn it, provided you're interested in doing so.

A former coworker, shy and conciliatory by nature, got a job at the most political, aggressive company in Silicon Valley. At the time, she told me, "The one thing I'm going to learn on this job is to be assertive." Some years later, she'd obviously succeeded, and I asked her how she did it. She replied:

> By dealing with people who were very assertive, I had to become assertive myself. I learned that I had to say no to people and not worry about what they felt, because otherwise they were going to tell me what to do and it's going to have an effect on me . . . If you look at how other people do things—how the engineers do things—they might be working 80 hours a

week, and they'll have those kinds of expectations of you. I have to be asser-tive enough to set my own limits and say no, this is what I need.

SUMMING IT UP

This chapter described the kind of person who becomes a tech writer and the kinds of backgrounds today's tech writers have. The next chapter explores different kinds of technical writing and educational opportuni-ties that can prepare you for your chosen field.

▪ ▪

[1] Kenneth Cook, Jr. and William Stolgitis, "Profile 88—Survey of STC Membership," *Technical Communication,* Journal for the Society for Technical Communication, First Quarter 1989, p. 40.

[2] Richard Bolles, *What Color Is Your Parachute?* (Berkeley, Calif.: Ten Speed Press, 1991).

[3] Barry and Linda Gale, *Discover Your High Tech Talents: The National Career Aptitude System and High-Tech Career Directory* (New York: Simon and Schuster, 1984).

[4] Bolles, *Parachute* (1978 edition), p. 90.

4
∎ ∎

GET READY

When you first bought your VCR—or microwave or other tech nological toy—and took it out of the box, did you read the unpacking instructions? Did you wonder who the writer was? Did you curse or praise him? Technical writing is so much a part of our lives that we take it for granted. Technical writers write everything from scripts for sales videos to the directions and warnings that come with an over-the-counter drug. This diversity presents a confusing array of choices to those considering a tech writing career.

What technology will you specialize in? This chapter will give you an idea of where to begin to look. Once you decide on a technical field, you're faced with the perplexing variety of media that technical communicators use to express concepts and describe products. Will you write articles, proposals, or film scripts? Or will you design presentations requiring few words and many visuals? This chapter describes four major categories of technical writing and the communication media they use. It goes on to describe educational programs and other ways to acquire the knowledge you might still need to enter your chosen field.

CHOOSING A FIELD

You can choose a field or let it choose you. I fell into writing about data communications software—my primary subject area—because the start-up company that first hired me happened to produce data comm software products. My primary career objective was to make a decent living as a writer, and writing was my main skill. The fact that I was a sophisticated computer user helped get my foot in the door.

If you are primarily a strong writer, you will choose a technical field based on either the available job market or a technical interest you already have—say computers or conservation. If you have some technical expertise, it will guide you toward a field where your current skills will help you begin. You may already be working in a technical field where you can make a lateral move into a writing position. More about that in the chapter "Breaking In."

Many writers switch between fields. They feel that tech writing skills, once acquired, are applicable to a variety of technologies. You can leave several options open, by learning the vocabulary and procedures of more than one technology and by becoming proficient in the communication media they use.

While technologies are varied and proliferating, most of the hiring is done by just a few industries:

- Computers
- Electronics
- Biotechnology
- Scientific research and development
- Aerospace
- Automotive
- Engineering and construction

Within these broad categories are numerous subfields. For example, computer software alone includes such disparate subjects as accounting, medical imaging, and engineering design. The next chapter, "Breaking In," presents ways to discover which industries are hiring in your geographic area.

In addition to deciding on a particular field, think about whether you want to work for private industry or for government. Each provides a very different work environment and places different demands on writers. For example, in private industry you'll participate in decisions about the writing style and design of documents and you're more likely to enjoy flexible work hours.

In contrast, a job in government, or on a government contract, is circumscribed by restrictions. One writer describes his first tech writing job in an aerospace company as "quasi military. Everything had to be kept in a safe because you're dealing with security documents. If you forgot to put it in the safe, you got a penalty. If you forgot three times, you got additional penalties. There was either too much work or no work. You spent a lot of time doing mindless things. The milspec format was very restrictive." This writer has a Ph.D. in English and was initially unfamiliar with both government requirements and the company's technology. If you have a military background and intense interest in aerospace technology, you might find this kind of job rewarding.

The military budgets huge sums of money for documentation—over $1 billion to document the space station alone, according to William K. Horton in *Designing & Writing Online Documentation*. And the volume of military documentation is staggering. Says Horton, "World War II fighter aircraft got by with 1,000 pages of manuals and drawings . . . Today, the B1 bomber has about 1 million pages of documentation."[1]

MARKETING OR JOURNALISM? THE MEDIUM IS THE MESSAGE

Writing is the heart of your craft. But the kind of writing you do is shaped by your medium. If you design a video, you cannot do the kind of writing you would for an environmental impact report or your audience will soon be snoring. Similarly, the purpose of the writing—whether it's to sell, instruct, inform, or entertain—will shape your style. If you're a technical journalist for a consumer publication, you'll use an informal, direct "voice" and cram your page with facts. If you're writing military specifications, you'll organize and express information following rigid guidelines. Each kind of technical writing places unique demands on a writer and requires different skills. The following sections describe a few purposes technical writing fulfills and the forms it takes. They're not meant to be comprehensive; an adequate description of any of these categories could fill a book!

MARKETING COMMUNICATION AND SUPPORT

Marketing communication is geared to sell. Its form varies with its audience. For example in the electronics industry, marketing produces data sheets—glossy, one-page handouts with a photograph or drawing of the product and a terse description of its functions. While the writing is often dry, the graphic impact must be "sexy"—very sophisticated and appealing.

In the pharmaceutical industry, marketing support materials appeal to doctors, who are less likely to respond to glitz. One of the largest pharmaceutical firms in California's Silicon Valley takes a low-key approach by giving doctors complimentary slide sets they can use to educate their colleagues and patients about medical and pharmaceutical matters. "It supports your marketing effort," says one of their pharmaceutical writers. "It spreads your name. Somebody receives this nice slide set and they're likely to remember who gave it to them."

If you write for marketing, you must know who the customer is and what the competition is doing. You will describe your product's strengths to convince customers to buy it. You will probably work in a wide variety of media, including overhead transparencies, slides, film, press releases, and magazine ads. You will work closely with artists and the production people who photograph or lay out the finished piece.

TECHNICAL MANUALS AND SPECIFICATIONS

Technical manuals and specifications contain the most complete technical information about a product, including its limitations. These publications range from simple operating instructions to extremely technical engineering specifications. Sometimes the audience consists of customers who are new to the kind of product you are describing. Often the audience consists of engineers or scientists.

While these documents are sometimes long, dry, and complicated, they are where the high-tech action is and where the jobs are. By writing technical manuals and specifications, you'll receive a free education in the latest technical developments in your field and you'll gain most of the skills you need to write less arduous publications.

Military specifications, which meticulously describe equipment produced for military use, fall into this category. Aerospace proposals do, too. "These are lengthy, detailed, complicated documents," says Dirk van Nouhuys, who started technical writing 29 years ago and now consults:

> The proposal for the C5A transport was shipped in a truck—one copy was shipped in a truck! [It filled boxes and boxes.] They're written by a special team put together for that purpose, which includes writers and editors and technical people. It's a terrible crash deadline situation. Some of these things—quite complicated ones—are written in six weeks. It's a bizarre Dionysian thing—people sleeping on the floor days on end and all that kind of stuff. It's still going on. Hundreds of thousands of people are employed doing it as we sit here.

So you can see that your choice of medium can have a serious impact on the quality of your life!

Tech writing has its horror stories. However, most technical manuals and specifications are produced in a standard 40-hour week, plus an occasional weekend before the deadline.

If you write technical manuals or specifications, you'll constantly learn more about your chosen technology. You'll work closely with technical experts. You will read inscrutable documents written by said experts, ask lots of questions, and organize lots of complex information. You will usually write to a fairly formal style, dictated by the publications department. For military specifications ("milspec"), you will follow government style guidelines.

If you write manuals, you'll often use desktop publishing software to produce a finished document. If your product is a computer program, you might write help information the customer will read on the screen. *Computer literacy*—the ability to use computer software and operating systems, and some understanding of how they work—is a must.

TRAINING MATERIALS

A training writer is an educator even more than other technical writers. In training, you'll produce course materials, including study guides, instructors' manuals, overhead visuals, test questions, and score sheets. Your materials will help train both company personnel who service customers and the customers themselves. For example, once a data communications company sells computer networks to several large companies, it provides training classes for the customers' network managers to teach them how to operate the network.

Computer tutorials are another form of training, which is usually packaged with computer software. Tutorials provide an interactive form of learning, in which the customer responds to questions on the screen or asks for more information by pressing a key or button. Computer tutorials are sometimes written in hypertext, a kind of data-base software that lets the user choose how to browse through information, rather than follow a rigid sequence. Tutorials usually use lots of graphic images, and some incorporate animation, sound effects, music, and voice communication.

If you write training material, you'll need skill at lesson planning and experience with the communication medium you'll use. You might work closely with trainers, computer programmers, graphic artists, or animators. If you have a teaching background, it will help you break into this kind of technical writing.

TECHNICAL JOURNALISM

Technical journalists write for several different audiences. For example, they can specialize in ghost writing books, lectures, and journal articles bearing the bylines of well known scientists or in writing product reviews for consumer magazines. The one thing these kinds of writing have in common is that their primary purpose is to inform, rather than train or sell. While ghost writers might write for an audience of scientists, most technical journalists need to translate highly technical concepts for a more lay public. The reader is usually there by choice, not because she bought the product and needs to find out how it works. If your article is boring, she'll flip to another page.

Ghost writing for scientists is much like manual writing—You will translate technical concepts into readable English for a specific audience and publication style.

Consumer journalism allows a more flexible, conversational style. Because consumer journalism is a glamour field, you will compete harder for a job and might make less money than your counterparts in the tech-manual trenches. If you like playing with technical toys, this field could be for you. For example, as a staff journalist for *PC World*, you might spend your time in a lab, testing competing software applications and writing about your findings.

A degree or experience in journalism will help you succeed in this kind of technical writing.

 ## ACQUIRING THE KNOWLEDGE YOU NEED

If you're a career changer with a degree, you'll learn the necessary skills for technical writing in different ways from someone still planning a college education. If you are an undergraduate deciding on a major, the following sections will help you with this decision. If you already have a degree or work experience, skip to the section "If You Already Have a Degree" for information on ways to gain additional knowledge. The next chapter, "Breaking In," will help you make the most of your current skills in a job interview.

IF YOU'RE CONSIDERING A DEGREE

If you don't yet have a degree, a technical communication major is one option to consider. One advantage of many technical communication programs is an internship opportunity—a low-paid, temporary junior position, which can lead to your first real technical writing job. Appendix A lists some of the colleges offering internships.

A recent survey of 72 organizations in diverse technical fields showed hiring managers prefer new writers to have technical communication degrees above other majors.[2] That sample was small, however, and English nearly tied with technical communication as the preferred major. Managers I interviewed did not have a preference for technical communication degrees.

Says Dirk van Nouhuys, who has been a technical publications manager for several companies:

> I don't think managers weigh tech writing degrees very heavily. When you're trying to get a job, there's a difference between dealing with a publications manager and somebody who's not a publications manager but who happens to manage a group that needs a technical writer. And I think that latter group—'cause those people tend to be degree oriented because specialized degrees are very important to them—is more likely to give weight to a degree in technical writing. But I don't. I know a bunch of publications managers and I don't know any publications manager who is explicitly looking for people with degrees in technical writing. I've seen some people who can't write who have degrees in technical writing and I'm so far not impressed with them. But it's not a handicap to have one.

Dirk's comments make sense when you realize that most hiring managers with technical writing backgrounds did not come from a technical communication curriculum. And managers, by their own admission, tend to hire people like themselves. However, many writing jobs are offered by people with engineering and business backgrounds, who feel less at ease hiring writers and are reassured by a technical communication degree. Also, the growing academic community in this field is promoting the desirability of technical communication degrees.

In deciding the kind of degree to aim for, consider not only what employers think but more importantly what will prepare you to do well as a technical writer. The skills you need can be gleaned from many different combinations of course work and practical experience. You can use the recommendations under "Academic Programs in Technical Communication" later in this section to guide you.

Also, there is life after technical writing, and you may want to move on to a career in business or some other field where a different mix of course work or a less-specialized degree will have prepared you better than a technical communication major. A narrowly focused major can open some doors but close many others.

Finally, let your unique set of interests lead you toward the curriculum you're most enthusiastic about. Your interests will guide you to the areas in which you perform best.

Opinions from the Field

When I asked what background they look for in technical writers, managers differed.

Carre Mirzadeh, publications section manager at Tandem Computers, reports:

> Our job descriptions say "degree in computer science, English, journalism, technical writing, or equivalent." My experience has led me to believe that a person from the humanities side or from the scientific or technical side can be a successful technical writer. However, I don't believe that a person with a technical background who's interested in writing will necessarily be able to become a successful technical writer. You need something more than an interest in becoming a writer.

When I asked her what that was, Carre talked about interpersonal skills and flexibility.

Says Dirk van Nouhuys:

> My favorite educational background is journalism and that is because people in journalism know how to ask questions, they know how to deal with getting information from people, they know how to write, and they have a clear image of an audience.

John Huber, technical communications manager at Apple Computer, says:

> A teaching background is helpful because in many types of technical writing, you have to organize material the same way you would organize a classroom presentation, starting with what the class—or the reader—already knows and building on that.

Biomedical writer Dan Liberthson reports:

> People in the medical and scientific disciplines have a bias toward hiring people with science backgrounds and science degrees. There are some industries, like bioengineering, in which I'd say you almost have to have a masters or a Ph.D. in science in a very specific area, like molecular biology. Extremely technical fields. I can't say whether a person with a general humanities back-

ground could understand them or not. I suspect they could, but the bias is to say that you need a scientific background.

Dan has a Ph.D. in English and is a successful biomedical writer.

Academic Programs in Technical Communication

In the mid-'80s, there were well over a hundred academic programs in technical communication, according to the 1985 edition of the STC publication, *Academic Programs in Technical Communication*. (To buy the current edition, write the Society for Technical Communication at the address listed in Appendix B.) This publication lists the 56 of these that replied to the authors' survey. Their figures indicate that the number of programs more than tripled in the nine years since their first edition—an indication of the rapid growth in this field.

A closer look at the descriptions of these courses reveals a wide range of focuses, some of them less than practical. One program titled "Technical and Professional Writing" provided journalism and English courses, with no technical requirement or practical experience. Another program not only provided writing courses directly related to technical industries, but also required a technical or science minor, some graphics and design training, and an internship in the field. This would seem to prepare you much more thoroughly for the realities of the tech writing field.

In a recent survey, 124 members of the Association of Teachers of Technical Writing were asked which of 48 undergraduate courses in 37 tech writing programs they considered most important. They ranked technical writing highest, followed by technical editing, graphics and design, an internship or research project, business and industrial report writing, and lastly, interviewing. (I would rate interviewing and interpersonal skills a lot higher than this.) The author concluded that "Overall, the results of this study indicate that none of the 37 undergraduate programs offer a complete list of courses that would be classified as the most important."[3]

If you are considering an academic program in technical writing, try to talk to graduates from the program who are working in the field. Ask them how well their education prepared them for their careers.

▮ ▮

Checklist 4–1.
Technical Communication Curricula

Below are characteristics you should look for in an academic program in technical writing:

1. Several writing courses, equivalent to at least a minor in English or journalism, with a minimum of one course in your chosen field (for example, science journalism or medical documentation).
2. Several courses in your chosen science or technology.
3. Courses or labs in word processing and computer applications. Even if you don't work in the computer industry, you must be computer literate, because in your tech writing career you will use computer word processing to produce documents.
4. An opportunity to learn interpersonal communication skills in a real setting, for example through a journalism course requiring you to perform interviews.
5. An internship in the technical writing field, with opportunities to attend meetings and experience the dynamics of a technical publications department in your chosen environment.

▮ ▮

The five characteristics listed in the preceding checklist are probably essential elements of a tech writing program. You would also benefit from a course in logic or rhetoric, a course in graphics or design, and courses teaching the production techniques involved in your chosen medium, be it paper documents, video, or interactive computer software. An education course teaching lesson planning would also be helpful.

If I had it to do over again, I would seek the educational opportunities I've just described. When I entered the tech writing profession, I already had my degree—a major in fine art, a minor in English, and a teaching credential. So I'm in the group I'll talk about next.

IF YOU ALREADY HAVE A DEGREE

If you already have a degree or some years of related work experience, you may be able to get your foot in the door without further study. The chapter "Breaking In" discusses how to do this in some detail.

With most degrees, you probably still need more technical knowledge to become a tech writer in a given slot. If you have a liberal arts background, or a science background from an unrelated field, you still need to learn product-specific information. You can learn this kind of information by taking a class, or through a company training program once you're hired.

If you have a primarily technical background, you can develop writing abilities that will bridge the gap between your understanding and that of your more technically naive reader. (Later chapters tell you how.) If you learn better from people than books, consider taking a writing or editing class to fill out your knowledge.

Taking Classes

Before signing up for a class, watch for one pitfall—many people think they need to go to school or take one more class before they go for their real goal. Most real goals are scary, and school is a handy way to put off pursuing them. If you already have strong writing or technical skills, you may have everything you need to get hired. And often companies will train writers.

On the other hand, classes can help you in several ways. One reason to take a class is to show prospective employers that you're up on the latest technological advances. Hiring managers admit they look for buzzwords on resumes. Buzzwords are product names and technical terms that indicate very specific experience.

Thus, if you want a job in a particular field, take a night course in its latest developments. For example, if you were applying for a job as a computer software writer in the late '80s, a course in the C programming language or UNIX operating system looked very good on your resume. Such trends change fast, and tomorrow a different buzzword will attract the hiring manager's eye.

Classes in either writing or technology can give you knowledge and confidence while you're pursuing your goal. Colleges and universities offer writing and technical classes you can attend in the evening or in summer school. If you don't live in a college town, phone your state university for information on summer school opportunities or extension classes you can take by mail.

In high-tech regions, you'll find a dizzying variety of course offerings. For example, one community college in California's Silicon Valley lists

- Business Communications
- Concepts/Electronic Office
- Business Computer Systems
- Introduction to Desktop Publishing
- Introduction to Telecommunications

and fifteen microcomputer classes, ranging from using a microcomputer to understanding its hardware. This college also offers a dozen computer programming classes, as well as classes in physics, ophthalmics, chemistry, and engineering.

If you need writing skills, take a class in expository writing, business writing, technical writing, or copy editing. I took a night course in copy editing at the University of California, Berkeley Extension, taught by a master editor, Max Knight. Not only did he cover the mechanics of editing, but he spent most of the course illuminating the principles of good writing. This he did with great clarity and enthusiasm. Take an editing course. You might get lucky, as I did.

Learning on the Job

Once you land your first job, you can acquire technical knowledge both on the job and through outside classes. I did both. My first technical writing job at a start-up company allowed me to come up to speed informally, by talking to the programmers in the cubicles next to mine and by attending weekly lunch meetings in which a staff programmer would present some technical topic. I also took the company's training course designed to teach customers about the product, a small computer network. This course allowed me to see how the training department presented the same material I would write about to the same audience I would address.

I next worked for a larger networking company, which offered week-long courses in data processing, data communications, and the company's networking products. I took as many of these as I could. Later, I attended a Pascal computer programming course at night at a community college. Lastly, I continue to build my own reference library of technical material about my strongest technical area, data communications. And of course, there's an ongoing educational process that occurs every time I research a new product and interview a technical expert.

Other writers report that they do a lot of self-directed study. Says Dan Liberthson:

> I went out, picked up a physiology book, and studied it as though I were taking a physiology course. I picked up an anatomy book. I studied it as though I were taking an anatomy course . . . I outlined the books and learned them—basically, memorized them. So I gave myself a course. And I found that physiology, as far as the pharmaceutical industry was concerned, was the most important course because it gave me a basic grasp of the anatomy and processes of the human body.

The other thing you have to do is read the journals. That's very important because doctors often talk in code, as engineers talk in code. There are certain terms that they use that are abbreviations, that are current terms, and if you don't read the journals and the medical magazines, you're not up on things.

Whatever your background, if you are an enthusiastic learner, know how to communicate with people, and enjoy writing, you can develop the skills needed to be a tech writer.

 SUMMING IT UP

This chapter described major areas of technical writing and talked about some ways you can acquire knowledge you may need to enter the area of your choice. In the next chapter, you'll learn ways to break in to your first, or next, technical writing job.

▪ ▪

[1] William K. Horton, *Designing & Writing Online Documentation* (New York: John Wiley & Sons, 1990), pp. 1–2.

[2] Dorothy Corner Amsden and Ann Parker, "Up the Ladder or Off the Track: Career Paths for Technical Communicators," 37th *ITCC Proceedings*, May 20–23, 1990, Santa Clara, Calif., p. CC-27.

[3] Earl E. McDowell, "Survey of Undergraduate Technical Communication Programs and Courses in the United States," 37th *ITCC Proceedings*, May 20–23, 1990, Santa Clara, Calif., p. ET-112.

▪▪▪▪▪▪▪▪▪▪▪▪▪▪▪▪▪▪▪▪▪▪▪▪▪▪▪▪▪▪▪▪▪▪▪▪▪▪

5

BREAKING IN

"What do I look for in a resume?" says Dirk van Nouhuys. "One thing I look for—and I'm somewhat embarrassed to admit this—is closely related experience to the job. Now the reason I'm embarrassed is that I know perfectly well that any smart person can learn very quickly quite a wide range of things that are not in their experience. But when you read resumes, you have to pick out one from another somehow, and that's one way to pick out one from another."

Everybody knows the job seeker's paradox—You've got to have experience to get a job and you've got to have a job to get experience. A tech writing job is a perfect example of this.

Classified ads for tech writers in both Boston and California's Silicon Valley papers frequently require three to five years of experience and, in addition, require experience in either the company's technology or computerized publishing system. What's a person to do?

▮ THE CHILL OF THE HUNT

A woman phoned me this morning seeking advice on how to break in to technical writing. She's had 15 years experience as a journalist, including a position as editor of a corporate newsletter, and she knows how to use several word processing programs and microcomputers. She's extremely well qualified for a technical editing or junior technical writing position. Yet her first few contacts with potential employers left her with a cold, unwanted feeling.

This phone conversation reminded me how hard it is to be a job hunter, particularly breaking into a new field. It reminded me to remind you upfront that self-confidence, perseverance, and luck are essential in any job search. Bolster your self-confidence and courage by any means necessary. Some job hunting guides can help you with this—Bolles' *What Color Is Your Parachute?* mentioned earlier, and Richard Lathrop's *Who's Hiring Who* (listed in the bibliography) both provide excellent guidance through the harrowing process of looking for work.

The rest of this chapter focuses on the particular requirements of the technical writing field and gives you some ways to improve your luck there. You'll find suggestions from managers on how to break in, and guidance on resumes and writing-sample preparation. You'll learn where the jobs are geographically and how to arrange and survive an interview.

 ## SOME SUGGESTIONS FROM MANAGERS

Hiring managers I asked recommended three ways to break into technical writing:

- a lateral transfer within a company where you are already employed and have shown some writing ability.
- an internship placement through an academic program in technical communication.
- a junior writer position in a company too small to afford an experienced writer, or in a company large enough to mentor an inexperienced one.

For career changers—particularly those with technical jobs, such as lab technician or product tester—a lateral move is probably the best way to break into technical writing. Says Tandem writing manager Carre Mirzadeh about how she broke into technical writing:

> I worked for a company that needed some writing done and there wasn't anyone else to do it, so I began to do it and once I started, I realized that I enjoyed it. When I first started writing, I didn't do it full time. I did it in conjunction with some QA (quality assurance) testing and tech support . . . Writing was initially maybe 10 or 20% of my job. And as the company changed, we hired people to take over some of the functions I was doing, and I spent more and more time writing. After the first year-and-a-half to two years, I was writing full time.

If you are already working for a company that needs a technical writer—whether or not they know they need one—you can volunteer. Think about what kinds of documents could enhance your current job; then write them. For example, if you deal with customers who repeatedly telephone with the same questions, write a question-and-answer sheet and send it to them. This sheet could save you telephone time and later help prove your writing abilities. Even if your writing efforts don't work into a full-time position, as they did for Carre, your question-and-answer sheet and other product-related pieces will provide writing samples to start your portfolio.

Another way to break in, mentioned by managers, is through an internship within an academic program in technical communication. If you are an undergraduate, this is one good reason to choose a technical communication major—your school will get you your first job.

Consultant Daunna Minnich hired new writers through a college internship program when she needed help. "I had some extra training to do but I figured I'd rather train somebody up the way I wanted than try to untrain somebody. Who wants to write unpacking instructions and things that are one to four pages? A junior tech writer." Thus the importance of an internship as part of a tech writing program! Appendix A lists some of the four-year colleges and universities that provide internships.

Internships are offered not only by degree-granting programs, but also by some colleges granting certificates or associate (two-year) degrees in technical writing. Internships have been offered through two-year programs at Austin Community College in Texas, Clark College in Washington, and Rock Valley College in Illinois. The Society for Technical Communication (STC) publication *Academic Programs in Technical Communications* lists more. (For the most recent edition, write to the STC at the address listed in Appendix B.) If you are interested in technical writing but have an unrelated degree and no related experience, consider getting a technical writing certificate in a program that offers an internship.

A third way to break in is to sell yourself "cheap" as a junior writer or editor. Companies that might hire junior writers include small companies that cannot afford to hire experienced writers and large companies with a layered publications department set up to mentor less experienced writers.

To break into a junior writer position, you need either academic or job experience in some writing field or in the technical field you will document. You also need to be aggressive about getting your foot in the door and you may have to make some "cold calls." More about cold calls later in this chapter.

All working technical writers I've asked have told unique stories about how they broke in. While no map emerged of a straight path to a tech writing career, their stories indicate that you too can find your own way. Who do you know who might help you? What unique skill do you have to offer?

 ## RESUME PREPARATION FOR WRITERS

A resume is a selling tool. That's all it is. This section focuses on resumes for writers, but no section on resumes can ignore these basic principles:

- A resume should contain only those details about your education and work experience that will spark the reader's interest enough to get you the interview.
- A resume should be as short as possible. "Approximately one interview is granted for every 245 resumes received," say David Hizer and Arthur Rosenberg in *The Resume Handbook*.[1] While this probably refers to personnel departments (a good reason to avoid them), your short resume assures the reader you value her time.
- A resume should contain no information that will elicit negative associations. Nothing about unusual hobbies, political or religious affiliations, age, divorce, sickness, or nonconformity in any form. You can always reveal your personal quirks in the interview, should the interviewer seem the sort of person who'd be positively swayed by such revelations.
- A resume should stand out. If you positively highlight your skills and experience, through thoughtful wording and typographic emphasis, and then format the information as elegantly as your tools will allow, your resume will stand out.

If you're writing your first resume, consult a resume-writing guide for details about the building blocks and principles of resume writing. The job hunting guides mentioned earlier also give good advice about how to present yourself on paper. Both resume-writing and job-hunting guides are listing in the bibliography.

A resume has two aspects: content and form. Think of them separately and polish each to a fine sheen.

CONTENT

Skills and experience provide the content of a resume. And, as Dirk mentioned at the beginning of this chapter, managers look for buzzwords. That means they prefer to hire people who've had hands-on experience with their company's specific technology and publishing system. This does not necessarily mean lots of experience.

If you know basic word processing, you can familiarize yourself with "hot" publishing packages by spending a few hours on a friend's computer. It helps if you're familiar with both the IBM and Apple lines of personal computers. If you're basically computer conversant, a few hours using a word-processing program or a computer system should give you license to add it to your resume. Even somebody who has spent months using a system forgets how after not using it for a while and needs to brush up upon returning to it.

In a separate section on your resume, include every technical skill you can think of that might spark the hiring manager's interest or meet a requirement in the company's official job description. Use several sections if you need to—one for the publishing tools you've used, another for equipment you're familiar with, a third for products you've documented. Brevity is paramount. Just list the buzzwords. Don't add amount of time spent with them, unless it's directly relevant to your job goal.

FORM

For writers, a resume is not just a resume. It's a writing sample. Therefore, more than for any other career goal, a writer's resume must be well organized, clearly written, and meticulously edited and proofread. "The first thing I look for is typos," says Sheila Borders. "If they've got a typo or a bad grammatical construction or just a poor way of expressing their job assignments, then I figure they're not exactly what I'd call a skilled writer."

Use active verbs and cut all words that don't contribute to your goal. A traditional resume is basically a list and should follow good list form— items are of parallel structure; verb tense and form are consistent; items are of similar length, although recent jobs can take more space than past jobs. If you are unclear about any of these principles, consult the chapter "Writing Is the Heart of Your Craft."

In addition to being well written, your resume has to look good. "I look for something that's got a fair amount of white space," says Sheila, "with type at least ten points, or bigger, because my eyes are getting old. I like it to be very dark type on white or buff paper."

When consultant Daunna Minnich was at Apple Computer, she received around 50 resumes a week, even when there was no job opening. Says Daunna:

> If somebody couldn't give me an attractive resume, I figured they weren't going to be able to figure out how to make a page look decent when they were actually writing. It was very easy to go through 30 or 40 resumes and pick out five to take a look at. They weren't crowded. They probably were only one page but might have gone on to a second. They certainly didn't have to be typeset, but they shouldn't look like some kind of term paper or like somebody had three strikeovers 'cause they didn't have time. You know— when you make a typo on an old-fashioned typewriter, and you keep going back and hitting it several times trying to fix the letter? I didn't care if it was chronological or functional—I really didn't care—if the thing looked attractive, then I'd decide to read it.

If you are uncertain about your design skills, have a friend with design experience look over your resume and suggest ways to improve its appearance. A second pair of eyes can really help you create an attractive resume.

■ ■

Checklist 5 – 1.
Resumes

Use the checklist below to make sure you've created the best resume you can:

- Have you listed all your technical skills?
- Is the layout attractive?
- Have you used active verbs and cut unnecessary words?
- Are job descriptions consistent in verb tense and form?
- Are items parallel?
- Is the resume free of grammatical and punctuation errors?
- Is it free of spelling errors and typos?

■ ■

WRITING SAMPLES

I have a gorgeous writing sample that came out of a team-written book with a big budget. We had artists, production people, typesetters, and enough money for two-color illustrations throughout. The customers ended up absolutely hating this book. It did not have the information they wanted, and the information that was there was not organized for easy reference. It was a tough lesson for me. It taught me both about the importance of knowing my audience's needs and about the pitfalls of a team-written project. Nonetheless, this book remained one of my best samples for a number of years. Why? Because it's pretty. (Also, the short sample of my writing does not reveal the overall organizational problems of the book.)

Most interviewers flip through your sample and judge it by how it looks. They rarely look deeply at the usefulness of your information and organization, perhaps because they are just too busy.

What's the lesson? Certainly not that you should write badly. But do make your samples look as crisp and lovely as you can, given your available tools. Even if you have only a typewriter, or worse an old dot-matrix printer, you can format your text with wide margins and plenty of white space, so the page is not gray. Break your text up with headings, lists, graphs, and tables. For more details about formatting, refer to the chapter "Planning for Visual Impact," later in this book.

Proofread your writing samples rigorously. Small errors have a way of leaping from the page at a casual glance.

Show flexibility by including samples written in different styles, even if you must write additional pieces just for your portfolio. Include a sample written in third person and one in second person; include a piece directed at consumers and one at a more technically sophisticated audience. Make sure to point out in your cover letter and in the interview that you can write in a variety of styles, so that the hiring manager does not just glance at one piece and decide you only write that way. Be aware that almost every company has its own house style and let the hiring manager know you're willing to go with it (if you are). It's part of the business.

Some writers find they can't go along with house style. For example, military specification ("milspec") writing is not everybody's cup of tea. You can determine this in the interview and later diplomatically remove yourself from the hiring process for that company. Of course, the luxury of such a choice depends on an abundant job market, where you stand a good chance of finding a more compatible house style at another company.

 ## WHERE ARE THE JOBS?

You're fortunate if you live in a pocket of technology, where high tech jobs are abundant. The big ones are listed on the facing page.

- California's Silicon Valley—the peninsula between San Francisco and San Jose.
- Boston around route 128.
- North Carolina's Research Triangle—Raleigh, Durham, and Charlotte.
- Pittsburgh, and the King of Prussia area near Philadelphia.
- the Dallas-Fort Worth area of Texas.

Other pockets include Portland, Oregon; Atlanta, Georgia; Salt Lake City, Utah; Minneapolis, Minnesota; and New York City. One way to improve your job-search luck is to be willing to look for work in one or more of these areas.

Even if your geographic area is not among the high-tech pockets listed here, it might support a prospering industry that hires technical writers. You can find out by looking through the yellow pages of your phone directory under the technical industries that interest you, by looking through the classified ads in your local paper, and by talking with career counselors. Even billboards along local highways can give you some clues about neighboring industries. For example, billboards along highway 101 between San Francisco and San Jose advertise computer and aerospace products.

 GETTING YOUR FOOT IN THE DOOR

It's all well and good to talk about who's hiring, but how do you get that first interview? How, when most advertised job descriptions require more or different experience than you have, are you going to get your foot in the door?

As job search books attest, the best way to find a job is not through the paper—it's through someone you know or through more creative search methods. According to Richard Lathrop, in *Who's Hiring Who,* "only 15% of job seekers find employment through formal job-publicizing systems." Such systems include public and private employment agencies, school placement services, and the newspaper. Lathrop reports 24% of job holders found their jobs by direct contact with employers and 48% through friends or relatives.[2]

Creative job search techniques for tech writers include a lateral transfer and an internship placement, described earlier in this chapter. Another big one is called networking.

NETWORKING

Networking means establishing friendly contacts with fellow professionals to share helpful work-related information. Networking through a professional organization is one of the best ways to find a technical writing job. For example, at the monthly meeting of the Silicon Valley chapter of the Society for Technical Communication (STC), hiring managers announce job openings, job seekers display their resumes at a table for that purpose and use the microphone to describe their availability and background, and a listing of current job openings is available.

Another professional organization, the San Francisco branch of the National Writers Union (NWU), provides a job hotline for technical writers who are members of the union, as well as monthly trade group meetings in which job leads are shared. Such organizations give you an opportunity to meet experienced writers and learn from them. In addition to job leads, you can find out what it's really like inside hiring organizations and what salaries they pay writers.

COLD CALLS

You are fortunate if your contacts provide you with plentiful interviews and eventually a job. But usually you will have to face the impersonal specter of the want ads or even the phone book yellow pages for possible openings. If you must resort to a "cold call"—phoning an employer when no one has referred you and you don't know the manager's name—here are some things to remember:

- *Find out as much as you can about the company before you call.* If it's a large company, ask their public relations department for their annual report, brochures, and recent press releases. If it's a smaller company, ask the telephone receptionist what kind of products they produce. Also ask people in your professional network for information they might have about the company. Once you know what kind of products they sell, seek out information—from acquaintances or the library—about those kinds of products. Who buys them? What are they used for? All this is so you'll appear knowledgeable when you make that cold call.
- *Even if you don't fit advertised job requirements, make that call.* In today's high-tech market, employers ask for the stars, knowing that they'll have to settle for the moon. If you can get through to the hiring manager with your sunny personality, obvious brilliance, or whatever other outstanding quality you offer, that manager will probably consider at least looking at your resume and perhaps a writing sample, even if you don't perfectly fit the company's job description. A follow-up call—from you—might lead to an interview.
- *Never go through the personnel department, if you can help it.* They know nothing about hiring writers and will screen you out im-

mediately on the basis of an official job description. When you call the company's main number, ask for the publications department (or training department or whatever department you intend to work for); then ask the person who answers the department phone for the name of the manager and ask if you can speak to him or her. If you need to explain your business, do so honestly. People in publications departments have stood where you stand and will usually be sympathetic and helpful.

THE INTERVIEW PROCESS

"If you depend on employers to lead the way in making a sound determination of your qualifications, chances are they will botch the job—and horribly," says Lathrop.[3] Be prepared to control the interview, by anticipating the hiring manager's needs and meeting them.

Unfortunately, the interview process is uncomfortable and complex for even seasoned applicants. The bottom line of interviews seems to be that you create a good impression, which you may be able to do in spite of your lack of experience and skills. The interviewer's subjective feeling about you can sway the result of the interview more than any other single factor. Says H. Anthony Medley in *Sweaty Palms: The Neglected Art of Being Interviewed*:

> The questions and answers of an interview are merely the tools used to make an evaluation, the trees in the forest of impression. Your relaxation, your confidence in yourself and your manner are far more important than the words you use in your answers.[4]

If you are new to the job search process, or even if you aren't, look at the job hunting books mentioned here (and listed in the bibliography) for tips on how to handle interview questions designed to throw you off guard, how to avoid making negative statements, how to deal with interview stress, and so on. The information in these books can help turn this often painful process into a challenging, successful experience.

WHAT HIRING MANAGERS WANT TO KNOW ABOUT YOU

Once you have some idea of what to expect from any employment interview, you can present your skills as a technical writer in their best light. Job hunting books suggest one key to a successful interview is to ascertain the hiring manager's needs and convince him or her that you can meet them. In this area you are fortunate. Technical publications managers all have a few simple needs in common. They want you to be able to ask questions assertively without making enemies, to understand the answers, to be "flexible" (a vague word but that's what they say), and to write. Let's see how these requirements affect an employment interview.

You Know How to Ask Questions

Asking questions is so critical to your performance as a tech writer that it's the first thing some managers look for in an interview. Says Sheila Borders:

> I interviewed someone recently who asked no questions at all about this particular position. And my immediate inclination was to think, is this the way they would interview a technical subject expert?

So ask questions. The Questions Checklist for Job Hunters, later in this chapter, suggests some questions you probably should ask.

You Can Understand the Answers

Managers want assurance from you that you're able to absorb and interpret technical information. They may use interview questions to find this out, or if they are not good interviewers, you'll have to volunteer evidence.

Some writing managers will ask an engineer to interview you and later ask that person how you responded to technical questions. Some managers will ask you technical questions themselves. Don't panic. You should know the buzzwords and general concepts of your technical area. But these interviewers don't expect you to know how many bits reside in the address segment of a particular data communication packet or to recite the exact structure of a particular DNA molecule. They're looking for signs of intelligence—for assurance that you'll understand these things with a minimum of tutelage. When you're asked a question you don't understand, ask one back. Continue asking until you understand the question and the answer. That's the exact process you'll use to get technical information on the job. If you do it well and learn quickly within the interview process, you'll impress all but a few literal minded stick-in-the-muds who want only a correct response.

Even if the interviewer doesn't ask technical questions, he or she still wants to know that you can handle technical subject matter. This is particularly true if you have a liberal arts background. Find an opportunity to describe how you researched a technical topic in school or on a previous job, or ask about the company's technical products and probe for details. Indicate that you're interested in technical wizardry, which you should be by now or you wouldn't still be considering this field!

You're Flexible

Flexibility, in this sense, does not mean that you practice Hatha Yoga daily, although that might help. Managers want you to be flexible in terms of their needs:

- You will not complain to your boss when your project gets canceled halfway through.
- You will not quit when you're assigned to work with the engineer writers call Attila the Hun.
- You will not freak out if the product schedule changes, forcing you to get the manual out by yesterday.

In short, you will not make problems for them. Technical writing managers tend to have very tight schedules and sometimes see a writer's need for problem-solving help as a demand for hand-holding—a pastime managers generally detest. Assure them in the interview that you enjoy working independently, that you like to take the initiative in solving problems, and so on. If you can do this without using such obvious cliches, you'll make an even better impression!

You Can Write

Be prepared to discuss your writing samples intelligently. Sheila Borders knows how to interview writers, so you can use her advice to reassure managers who don't know how:

> I usually want writing samples brought to the interview, so I get the person explaining about the book the samples came from. I want to know who the audience is. I want to know something about the project that it was developed under . . . What restrictions surrounded the production of the book?
> The physical presentation isn't necessarily the writer's choice. You see this grungy manuscript—8½ by 11—it looks like it was done on a typewriter. It doesn't look great in our environment where we're dealing with fairly fancy looking books. So your first impression is, is this all you can do? After I find out what the circumstances are—this was a start-up company with a typewriter—then I get past that physical, judging-the-book-by-its-cover idea and move into the content. I see how many questions I can come up with that don't seem to be covered. In a real snap judgment sort of way, I ask those questions—see how the writer responds to that quick critique.

One reason she does this is to make sure the applicant is truly the author of the sample. "I figure if they lived it," says Sheila, "there's no hesitation in their response."

In the interview, describe the restrictions under which your samples were written and point out each document's strengths, whatever they may be. Of course, mention any positive responses or comments the document drew from customers. And mention ways it might have helped the company, for example "This installation guide reduced service calls by 30%."

WHAT YOU NEED TO KNOW

While the employer is choosing a new writer, you are choosing where you'll spend most of your waking hours and energy. In the interview, you'll need to find out as much as you can to make this critical decision. Isn't it natural to ask questions?

Below is a checklist of specific questions you might ask in an interview for a technical writing position. Don't try to ask them all. Choose those questions that will serve your unique concerns and create questions of your own.

Checklist 5-2.
Questions to Ask a Prospective Employer

- Are your documents written in second or third person? Active or passive voice? May I see an example of the kind of writing you do in this department?
- How technically naive or sophisticated are your readers?
- Do you have a lab where writers can use the product? Can I see the product?
- How is the department structured? Would I report directly to you or would someone else monitor and evaluate my performance?
- It is possible to meet the writers I'd be working with?
- Is there an editorial staff? Who has the final say on edits—the editor or writer?
- Who reviews my documents? Who will be responsible for their final content?
- Are the engineers accustomed to working with writers? How cooperative are they?
- Will I be responsible for camera-ready copy?
- Where is the publications department in the organization chart?

Obviously some of the interviewer's answers to the questions in the preceding checklist might not please you. Try not to react in the interview. Weigh your impressions later to decide if the job's good qualities outweigh aspects you don't like.

Before the interview ends, ask when the interviewer will reach a hiring decision.

THE QUESTION OF MONEY

Most technical writers I interviewed told me their first tech writing job paid them twice what they'd been making at their previous job. Technical writing pays well. How well? You need to do this research for yourself. Salary surveys, published by organizations like the STC, are usually way out of date by the time they reach you.

On salary matters, networking is of great value. I determine my rates by periodically asking other writers how much they make. If they don't want to tell me, that's fine. But most do, and they ask the same in return. At meetings of your chosen professional organization, you can ask the question of several writers and get a sense of what writers with different experience levels make and what different companies pay. For staff positions, salaries vary quite widely between companies.

If this is your first technical writing job, you may have to accept a salary low on the company's scale. Accept it gracefully, unless it's ridiculously low. And congratulate yourself for landing the job!

If you already have an entry-level position and are going for your next job, you have more leverage and can set a higher goal. For help negotiating pay within the interview, refer to the job hunting guides listed in the bibliography.

 FOLLOW-UP

A hiring manager told me she'd recently interviewed two applicants with equally impressive credentials and was at a loss over how to choose one. Which applicant did she finally hire? The one who sent her a thank-you note after the interview. Can you afford to skip this step? Obviously not.

After the interview, send a hand-written card thanking the hiring manager for his or her time. You'll create an even better impression if you mention a point he or she made.

During the interview, the interviewer told you when he or she would reach a hiring decision and probably reassured you that you'd be notified. Don't wait for this call. When the stated time has elapsed, phone and ask how the decision-making process is unfolding. If you receive a rejection at this point, have the courage to ask how the decision was reached. Don't take the answer too seriously—it's difficult for employers to give you a complete, honest answer as to why they rejected you. But if you ask nondefensively, you might get some feedback that you can use to improve your performance in future interviews.

Most important, don't get discouraged. Rejections are a necessary part of breaking in. And if that phone call yields an acceptance, it's time to celebrate!

 SUMMING IT UP

This chapter suggested ways to break into technical writing and provided tips on resumes, writing samples, and the job interview. The next chapter lets you glimpse inside the office window. What will you find yourself doing in a typical day as a technical writer?

■ ■

[1] David Hizer and Arthur Rosenberg, *The Resume Handbook* (Boston: Bob Adams, Inc., 1985), p. 20.

[2] Richard Lathrop, *Who's Hiring Who* (Berkeley, Calif.: Ten Speed Press, 1977).

[3] Lathrop, *Who's Hiring Who*.

[4] H. Anthony Medley, *Sweaty Palms: The Neglected Art of Being Interviewed* (Berkeley, Calif.: Ten Speed Press, 1984).

6

■ ■

A DAY IN THE LIFE

The workstation lab is a large, warehouse-like space. Fluorescent ceiling lights glow high above long tables, where programmers and hardware engineers sit absorbed with display screens or tangles of wire.

Ken sits on a high stool frowning at a computer display. He's unaware of me as I move to within a few feet of him. I always feel hesitant to interrupt an engineer in the throes of a deadline. But his deadline is also mine.

"Ken . . ."

He jumps at the sound of his name. "Oh, hi. I bet you want to know what I did to the interface."

"Yes," I tell him. "It needs to be in the manual."

"Well, if the software would stop flying off into the ozone, I might be able to show it to you!"

I sit on a stool looking over his shoulder for the next hour, while he "debugs the code" (figures out what parts of his computer program are causing the trouble) and intermittently tries to show me how it's supposed to work. I take notes and sketch a few screens, all the while doing my best not to disrupt his debugging process.

This scene from a day in my life illustrates one way technical writers spend their time. Ask a technical writer what he or she does in a typical day and you'll hear an extraordinarily varied list. Said one writer, "I don't have a typical day."

TECH WRITERS WRITE, RIGHT?

One thing is clear—tech writers spend most of their time at activities other than writing. One purpose of this book is to prepare you for some of the nonwriting tasks you'll perform, like researching a technical product in a corporate environment, where information is usually not neatly catalogued and communication is your primary research tool. Estimates of how much time a tech writer spends writing vary between 25 and 35%.

"We did a survey at Apple at one point about how much of a technical writer's time is actually spent sitting in front of his or her terminal writing, as opposed to being in meetings or interviewing people or greasing the skids in other ways," says Dirk van Nouhuys. They found that writing time amounted to about 25%. "But that was Apple at a certain stage," Dirk explains, "I think probably for the average technical writer it's somewhat higher—say 35%."

Other writers I've asked about the percentage of time they spend writing concur with Dirk. Says consultant Linda Lininger, "In-house writers tend to be bludgeoned with meetings. As a free-lancer, I avoid them. I'd say most in-house writers are lucky if they write a third of the time, because there are numerous interruptions and all those crazy meetings they go to."

Says another writer, "It's probably one third meetings, one third writing, one third other stuff—dealing with politics and personalities and office supplies and overhead of some sort."

These writers' estimates of writing time agree closely with the results of a survey of 25 tech writer graduates of a technical communication program at Michigan Technological University. The pie chart in Figure 6-1 illustrates how they spent their time.

 WHAT DO YOU DO BESIDES WRITE?

What you do in a day depends on the kind of technical writer you are and on the stage of your project. A training-film writer's day will be different from that of a technical journalist. Says a computer software writer:

> As a writer of books, in a typical day I might interview a couple of programmers, maybe with a tape recorder, to ask them what they meant by their

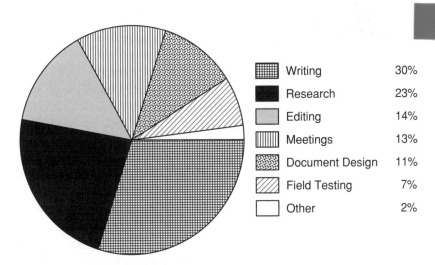

▦	Writing	30%
■	Research	23%
▨	Editing	14%
▥	Meetings	13%
▦	Document Design	11%
▨	Field Testing	7%
☐	Other	2%

Figure 6-1. On-the-job activities of 25 graduates of Michigan Tech's program in scientific and technical communication.[1]

obscure scribbles in a review copy of something I'd written; go to a staff meeting, where a manager tells us what work is on its way and how deadlines are shifting and so on; maybe go to a meeting of a project team; spend a couple of hours revising my current draft in the light of what I'd learned from the earlier interviews. Gee, there really is no one typical day. I've used up over eight hours already.

Says hardware writer and former sculptor Billie Levy about her day:

It'll be a combination of mashing away on the terminal, maybe getting up and looking at drawings, maybe making a drawing that I want to be in the book. I don't like sitting at a desk. I get restless and my neck hurts. So I go in and see what the technicians are doing, and the engineers, and I watch what they're doing, and I ask questions.

ADDITIONAL RESPONSIBILITIES

Besides undertaking the varied activities inherent in producing a technical document, writers enjoy additional responsibilities within a high-tech corporation. These can include:

- participating in product design decisions.
- writing the specification the technician or "implementer" will use to build the product.

- testing the product.
- editing the *user interface*—the part of a computer program that the customer sees.

As a technical writer, you will often participate in product team meetings, which include those representatives from the scientific, marketing, and testing departments that are directly involved with the product. At these meetings, team members decide scheduling, product design, packaging, and other issues that are constantly reexamined during the product development cycle. When should you join the product team? The earlier the better.

At the beginning of a project, you can act as the user's advocate, suggesting design modifications that make the product easier to use. In some companies, you'll produce the design specification, which the product implementers will use to build the *prototype* or *alpha release*—the first model of the product. In an article in the STC *Intercom* newsletter, Robert E. Erickson, a tech writer at a medical electronics firm, recommends the writer produce the specification. This spec can later evolve into the operator's manual:

> In our company, this has worked so well that it has become an established procedure for solidifying and communicating the 'big picture' to programmers and other system developers. The manual has evolved for use as a rallying point for argument and resolution of operations issues, before the design is committed to hardware or software.[2]

Erickson describes an unusually high level of writer involvement. This would certainly eliminate many of the problems writers tend to uncover later in the development process, when writer and product usually meet.

Some companies involve the writer only at the end of the product cycle, when design decisions have been finalized and the schedule is too rigid to permit unforeseen problems. Writers always uncover unforeseen problems! One of your most valuable nonwriterly roles is product tester.

You'll perform this role because you need to use the product to see how it works and to check it against the document-in-progress. In doing so, you will uncover and report functional failures (bugs) as well as features that are difficult to use. Most larger companies have established a bug reporting procedure and provide a "bug report" form you can fill out and submit. For the sake of your document and the customer, try to gain access to the first working version of the product.

If you are documenting a computer program, this is the stage at which you'll suggest rewording error messages, system messages, and other text that are part of the user interface. You should inform developers of any grammatical and punctuation errors, inconsistencies in use of terminology, terms and acronyms that need definition, and just plain incomprehensible stuff. Many developers speak English as a second language and design some rather surprising sentence structures into the user interface. Developers welcome a writer's input on these errors. But watch how you say it. Diplomacy is paramount.

 ## DIFFERENT STAGES, DIFFERENT DAYS

Different stages of a writing project demand different skills of a writer. When you're planning a project, you'll engage in a different set of activities than during the production part of a project. The following subsections list activities you might perform at different stages. Projects interweave, however, and you'll find yourself in the planning phase of one, while another needs last-minute changes. Says John Huber:

> While you're working on project B, you might be interrupted seventeen times with questions about project A, which you thought you finished three weeks ago, but the artist wants clarification on your request for a graphic, the editor wants to talk to you about a word you used, or whatever. While you're working on B, A keeps biting you on the ankle, and maybe the one before A, as well!

So be aware that some of the following activities lists are more tidily arranged than in the real world.

A RESEARCH DAY

- You read the internal reference specification ("spec") for a computer program and determine which part of the information will affect the user.
- You write a list of questions to ask the spec's author.
- You set up interviews with the product designer and implementer.
- You talk with the editor about "house style"—how is this type of manual organized? Can you see others that have been produced in the past?

A PLANNING DAY

- You ask your boss about the "product freeze," the date beyond which the product will not change. The boss sends you to find that out.

- You ask the engineering manager about the proposed freeze date; then write a memo to the manager and your boss summarizing the conversation.
- You attend a product team meeting, in which everyone reports on their schedule in relation to the product.
- You sit down at your computer and outline what you know about the product.
- You print your outline, then scribble notes all over it.

A "WRITING" DAY

- You talk with the artist about ideas for illustrations. She asks you to sketch your ideas.
- You sketch a group of microcomputers and draw arrows showing their relationships.
- You go to the engineering lab to look at a processor board. You make note of the number of cable attachments it has.
- Back at your desk, you sketch four more art ideas, number them as figures, and create a numbered list of proposed figures for your manual.
- You photocopy the list for the artist.
- You continue writing instructions on how to install the processor board.

A PRODUCTION DAY

- You insert index entries in your word-processed document file; then run a computerized index generator to create an alphabetized list of entries with page numbers.
- You send the list to the laser printer. When you go to the printer to retrieve it, you find your index list full of unintelligible characters.
- You phone tech support and ask to have someone look at the printer. The technician on the phone tells you how to "reload the printer fonts."
- You follow the instructions, which do not work, and then get a cup of coffee.
- You phone tech support again, and the technician comes over to fix the printer.
- You are late for a staff meeting.
- You stay overtime and finish your index.

Obviously, a tech writer uses many skills besides writing, as we already know from previous chapters. How many do you use on a typical day? Table 1 lists some of the activities described in the preceding lists, together with the skills they require.

Table 6-1. Skills Used in a "Typical" Day	
Activity	*Skills*
Reading specs and listing questions; writing an outline	Clear thinking; organizing ideas
Sharing a team project	Clear communication; active listening; negotiation
Ascertaining the project schedule	Time management; diplomacy; assertiveness
Writing a memo	Organizing ideas; clear writing; diplomacy
Planning figures	Visual thinking
Creating an index	Organizing ideas; computer literacy
Getting the printer fixed	Clear communication; computer literacy; persistence; diplomacy

 ## SUMMING IT UP

This chapter gave you a detailed picture of a day in the life of a technical writer and continued to describe the kinds of skills you need. The next section of this book tells you how to begin the documentation process, starting with concrete ways to research a technical product.

▪ ▪

[1] James R. Kalmbach, et al., "Education and Practice: A Survey of Graduates of a Technical Communication Program," reprinted with permission from *Technical Communication,* Journal of the Society for Technical Communication, First Quarter 1986, pp. 21–26.

[2] "Role of the Technical Writer on the Product Design Team," *STC Intercom,* December 1989, p.1.

PART TWO

GATHERING INFORMATION

7

KNOW YOUR SUBJECT

"Technical writing is probably about 70% research," says Billie Levy. "You just don't go in and sit down and start spitting out words. You can feel sort of guilty about this at first, but the research—looking at drawings or at the system and tracing down the air lines or the water lines—is legitimate. You can't do a good job of writing without it."

Billie's comments remind me of an ancient story about three blind men trying to understand an elephant. The first felt the trunk and said, "An elephant is like a long cylinder." The second patted its side and said, "The elephant is like a large wall." The third, stroking its tail, said, "No, it's much more like a snake." If you don't take the time to understand the product, your document might lack some vital details.

BECOMING A SUBJECT EXPERT

Some people think tech writers don't need to know much about their subject. They see the writer as a mechanical editor, cleaning up the grammar and punctuation and "prettying up" the format. Experienced writers think otherwise.

"The thing that I learned the most from my initial experiences is, whenever I come to a new project—before I even set finger to keyboard—I read *everything*," says consultant Linda Lininger. "And I didn't do that at first. I mean, I didn't know to do that. But now, if I am confronted with a new product, I read the manual, cover-to-cover."

Knowing as much as possible about your subject helps you in several ways. First, the more you know, the more technically accurate your document will be. Unlike other kinds of writing where verbal agility is paramount, technical writing demands accuracy first. You may be a superb stylist, but if your material isn't accurate, your readers will experience confusion and anger when they read your work.

Self-confidence is the second payoff for subject knowledge. You can press developers for information, because you're confident about the terminology and nature of the product. You can express your questions clearly and organize them effectively.

KNOWING YOUR PURPOSES

The first question to ask yourself, in researching a technical topic, is what is the purpose of the product? The second—what is the purpose of my document?

When you first receive your writing assignment, you'll probably get a brief description of what the product does. Think about this information. Try to imagine how customers will use the product and what they'll expect from it. What primary and secondary purposes does it fulfill? Approach the product from the customer's point of view. (Later, the chapter, "Know Your Audience," will help you refine your understanding of the user, so your writing will address the user's needs.)

Next, ask yourself, what is the purpose of the information I'm seeking? For example, if you're writing a marketing brochure about a computer program, your purpose is to convince the customer to buy the product. So, in your research, you'll ask questions about the program's features and how those features fill a customer's specific needs. Those questions will elicit information that serves your marketing purpose—convincing the customer.

On the other hand, if you're writing a tutorial about that same program, you'll need a different kind of information. Your purpose, in this case, is to teach the user how to use the program. You can step into your reader's place by learning how to use it yourself. You'll also need to know about the user's experience with computers and about the nature of the tasks the program helps the user perform. You'll look for information that serves your purpose—telling the user, step-by-step, how to use the product.

You can see that the kind of document you're creating shapes the kind of information you need.

 ## LEARNING ABOUT A TECHNICAL PRODUCT

You will learn about a technical product by

- reading technical documents.
- questioning technical experts.
- attending in-house training classes.
- using the product.
- thinking clearly about the data you've gathered.

Your most important sources of current product information are the technical people working on the product—its designers, implementers, testers, and quality assurance specialists. Because your communication with them is so important, the next chapter, "Communicating with Engineers," is devoted to that process. It also describes ways to create and order questions in preparation for the interview. The rest of this chapter explains how to research your subject before you approach these specialists.

GATHERING INFORMATION SOURCES

To learn about a technical product, first find out what sources of information are available. The Information Source List in this section can help you. Your manager will know some of the answers to these questions and can direct you to people who'll answer the rest.

At this point, also ask your manager for a preliminary list of people who will review your document and the names of their departments. Later, if you need information from marketing or product testing, you'll know whom to contact.

When you fill out the Information Source List, you'll probably leave some questions blank. For example, questions 1 through 3 ask for the names of the primary resource person for your document, the product designer, and the product implementer. If your company is small, these may all be the same person. In a large company, these roles might be filled by three different people, and you'll need to go through one to find out about the others.

It helps to have more than one person to contact with questions. Sometimes your primary technical resource won't be available or won't know the answers to all your questions. You can use your other contacts to get your answers.

Once you've filled out the Information Source List, you'll know where to find copies of product specifications and to whom to go for information.

∎ ∎

Information Source List

1. Who will be my primary resource (or first contact)?

Name_____Phone number _____

2. Who is the product designer (or second contact)?

Name_____Phone number_____

3. Who is the product implementer or developer (or third contact)?

Name_____Phone number_____

4. Does my company offer training classes I can attend on the use of this product? Name of class_____

Date offered_____Register with whom?_____

Phone number_____

5. What manuals are already available on this product?

_____

_____

6. What engineering documents (e.g., functional specifications and change notices) are available?

_____

_____

7. Is an operating version of the product available for me to use? Who can show me how?

Name_____Phone number_____

∎ ∎

READING TECHNICAL DOCUMENTS

To ask intelligent questions, you first must read available documents and identify where information is missing or incorrect. The checklist in this section can guide you. The subsections after the checklist provide details about the different kinds of documents you might read to research your topic, as well as tips on how to read them.

██

Checklist 7 – 1.
Identifying Missing Information

- Do procedures seem complete or are there steps missing?
- Does the sequence of steps seem logical?
- Is the information specific? "Developers occasionally write *is recommended* or *is suggested* when they mean *is required*," says Daniel Nolan in an article in *Technical Communication*, "and they write *should* or *may* when they mean *must* . . . When you see *is recommended, is suggested, should,* or *may* . . . you *should* probably ask questions." [1]
- What is the active agent? Often technical documents are written in passive voice, so the cause of action is unclear. For example, "a secondary station response must be initiated." Who or what is doing the initiating?
- Is the information current? Question numerical values to be certain they haven't changed since the spec or previous manual was written. Check version numbers, dimensions, ranges, limits, and default values.
- Are all technical terms defined? List all terms, along with their definitions, and question the ones you can't define.
- Are all acronyms spelled out? List all acronyms and question the ones you don't understand.

██

Keeping Track of Questions

Keep track of your questions in one or more of the following ways:

- Write them in a notebook that you reserve for information about the product.
- Write them on your copy of the spec or previous manual.
- Write them on sticky notes that you can attach to the spec or manual page in question.

I like the sticky notes, because I move them to the middle of the page when a question is answered, or I stick them to my computer display screen when I have trouble finding an answer.

Reading Product Specifications

The first step in researching a product is to read all available product specs. Specs differ between product types and between companies. For example, if you're writing a product monograph for a drug company, you won't find anything called a spec, but other documents serve the same function.

Says one pharmaceutical writer:

> The first thing you usually do is a library search through Medline. Medline is a data base. You design a literature search to pull out everything having to do with the drug. You look through your company's literature, the final reports on clinical trials, on toxicologic trials, and so forth.

If you write about computer software, you'll encounter

- Functional specifications
- Internal reference specifications
- External reference specifications
- User interface change notices
- Software change notices

and a host of other technical-sounding documents. You'll probably refer to them by acronyms—the ERS or the UICN—but for the purpose of this discussion, they're all "specs."

Specs have two uses to you as a writer: They are (1) a source of information about the product, and (2) a source of questions for the product developer (who may also be the author of the spec).

Some specs are extremely well-written and clear, but many are not. Spec writers usually don't define jargon or acronyms. They tend to use terminology inconsistently, calling something an "index unit" in one place and an "index object" in another. Therefore, some skill in reading specs will help you extract the information you need, without succumbing to confusion or boredom. The following tips should help.

Tips on Reading a Spec

1. Make a photocopy of the spec so you can mark it up.
2. Write your purpose at the top of the spec. What kind of information do you need from the spec? For example, "Installing the software."
3. Skim the whole spec and highlight everything you might need—draw a line along the side of relevant sections.
4. Go back and read the sections you've highlighted. As you read, do the following:

   - Mark or circle paragraphs that you can use verbatim in the first draft of your manual. (Plagiarism in tech writing is totally legit, as long as you "lift" material from within your company.)

- List terms and acronyms, as recommended earlier in this chapter.
- List questions that seem relevant to your purpose. For example, if the spec says, "The default of 300 applies for most configurations," you might ask, "Under what circumstances would the user need to use a different value?"

Reading Manuals

In addition to specs, other written materials can be sources of information and questions. Most commonly, you'll be able to read previous manuals about earlier versions of the product or about similar products. These are valuable not only for technical information but also as examples of the manual formats and "house style" used by your company.

Reading manuals is similar to reading specs, only easier. The following tips will help you read a manual for information and questions.

Tips on Reading a Previous Manual

1. Skim the manual and determine how much of it you can use.
2. If you can use a lot of it, ask for an "online" copy or a "soft" copy—one that you can manipulate on your computer. Later, you might be able to build your new manual using large chunks of existing documentation.
3. Try to get a copy of the manual that you can mark up. Or if you can get an online copy and it's not too long, print it out. Now you can mark up the copy as you did the product specification.
4. Highlight useful information.
5. Cross out information that you know is outdated.
6. In a notebook, list terms and acronyms next to their definitions, as described earlier. The acronyms and terms in a previous manual should already be defined.
7. List questions that you think need to be clarified to write your manual.
8. Talk to the previous manual's author, if available, to find out what problems and questions came up while documenting the last version of the product.

Sometimes no spec, previous manual, or written material is available about a product. If the product is a ground-breaking, one-of-a-kind new gizmo, you may have to go straight to its designer for information. However, most products are like something else that already exists, perhaps from a different company. To increase your knowledge of your subject, read manuals about competing products, which you can sometimes get from your marketing or engineering department. At least you'll understand some of the concepts that went into your product. This knowledge will help you when you talk to your product's developer.

ATTENDING IN-HOUSE TRAINING CLASSES

Many companies have training classes for customers and for field support personnel—the folks who go to the customer site to fix things. You will probably be permitted, if not encouraged, to attend. As I mentioned in the chapter "Get Ready," I recommend taking as many as possible.

During the research phase of your writing project, training classes can teach you generally about a technology and specifically about your company's products. They also help you make your terminology consistent with the terminology used by the training department.

Additionally, you can make valuable contacts within the training department—people who can answer your questions or review your work at a later date. The following tips will help you get all you can out of in-house training classes.

Tips on Using In-House Training

1. Ask the trainer if he or she is willing to be contacted later to answer questions about the product.
2. Ask if the trainer is willing to be on your list of reviewers. If so, add the trainer to your reviewers list, along with his or her mail stop and phone number.
3. During the class, ask lots of questions, particularly about the product characteristics you'll be documenting. Everyone in the class will benefit from your questions.
4. List terminology and one or two short sentences defining each term.
5. Ask the trainer to clarify any definitions you don't understand.
6. Ask the trainer to define acronyms, if he or she hasn't already.
7. Hoard all class handouts.
8. Later, go through class handouts and highlight any information that would help your reader understand the product. Remember, you can plagiarize from in-house documentation. Lift whatever is helpful.
9. File class handouts for later reference. For example, you can refer to them to see how the training department uses certain terms.

USING THE PRODUCT

Some people learn best by doing. If you are such a person, do everything you can to get your hands on a working model of the product. If you are not such a person, you should still use the product in its final phases, to make sure the developers have accurately communicated all its features and have not incorporated new features that they haven't told you about. This is critical. Most of the time you'll document a product that is not finalized or "frozen." Developers, creative folks that they are, will try to squeeze as many features as possible into the product between its inception and the rush-out-the-door. The only way you can check the final accuracy of your manual is to read it while you use the product.

If the product is too specialized or easily damaged for you to use, ask to observe it in use at a test site, in the quality assurance lab, or wherever your company puts the finishing touches on the product.

If you cannot possibly view the product's final operation, make at least one close contact among the product's testers (who may be called, among other titles, "field engineers," "test engineers," "quality assurance testers," or the product developer). Ask your contact to use your manual with the product and mark anything that works differently or appears differently than described.

If you can't check the manual against the final product, you might be writing fiction without even realizing it.

CLEAR THINKING

"Learning without thought is labor lost," said Confucius.

Clear thinking is an important part of research. You don't need a course in logic to become aware of your assumptions: Learn to distinguish between when you know something and when you are just guessing. You will do a lot of guessing in technical writing and will sometimes have to write information that you are not sure about. Be aware that you are guessing and *check your facts* later.

Sometimes writers don't distinguish between guesses and facts. They think, "It's obvious that the product must work the way I describe it here," without questioning the parts they've assumed. Such fuzzy thinking is usually the result of deadline pressure, shyness, or technophobia—three frequent enemies of clear thinking.

When you're under deadline pressure, you do experience reluctance to follow up on technical questions. How does the product really look? Does this old drawing still describe it, or should I go into the lab and see how it looks now? If I find out it looks different, I'll have to ask the artist to do a new drawing right away, and I might not make my deadline. Well, maybe this drawing is close enough.

Dealing with uncooperative developers can bring out latent shyness you never knew you had. How does the product work? Does it *replace* another device or *attach to* the other device? I'll ask Pat. Hmmm. Last time I interviewed Pat, she fumed, "Let's get this over with. My schedule is full." Maybe I'll just say the product attaches to the other device. That can't be too far off.

Technophobia goes something like this—My God, this thing is complex. I'll never figure out how it works. I can't even understand the spec enough to ask intelligent questions. I'll make a fool of myself with the developer. I know. I'll take his spec, add a few commas, break up the long paragraphs, and format it to be the manual.

So, pressure, shyness, and technophobia are enemies of clear thinking. How do you know when you're making false assumptions? Ask yourself *how do I know* what I know?:

- From studying similar products.
- Reading the product specification.
- Talking with the product developer.
- Using the product recently.
- Guessing.

If you need to guess to finish writing something, attach a note to the page and turn your guess into a question to ask the developer. For example:

"To include a user logo on the header page of a print job, the following eight lines must be placed [WHERE?] . . ."

Embedding questions in your draft is one way to avoid making assumptions.

SUMMING IT UP

This chapter described the importance of knowing your subject and summarized ways to research a technical product. It then provided tips on how to get the most out of your sources of information. Getting information from your most important source is described in the next chapter, "Communicating with Engineers."

▪ ▪

[1] "Analyzing Technical Input," *Technical Communication*, Journal of the Society for Technical Communication, Third Quarter 1990, p. 260.

8

■ ■

COMMUNICATING
WITH ENGINEERS

The richest source of current, accurate technical information about a product is the technical person working directly with it. In this chapter, that person is called an engineer, though he or she may be a chemist, computer programmer, or chief scientist. A critical part of your job is to communicate effectively with the engineer.

The New York radio talk-show host Barry Farber says you know real communication is happening "when everybody around, though fully dressed, seems to be sharing a hot tub."[1] That's a little farther than you need to go with an engineer, except under circumstances beyond the scope of this book. Your purpose is to establish real rapport. And that rapport serves an important purpose—getting information.

Technical writers and journalists have a lot in common. Both have to find things out quickly, by asking focused, well organized questions; both have to write concisely; and both have to meet deadlines. The interview skills that good journalists use work equally well for technical writers. These skills involve:

- choosing the authority most likely to have the desired information.
- carefully preparing questions to ask.
- effectively communicating during the interview.

- taking accurate notes.
- fleshing out notes after the interview.

PREPARING FOR THE INTERVIEW

The more prepared you are for an interview, the better it will go. If you know ahead of time what you will ask, who you are asking, and how you will use the information, you can relax and concentrate on making contact with this person, the engineer.

A story goes that a press conference was held to celebrate the rerelease of the movie *Gone With the Wind,* starring Vivien Leigh. And Vivien Leigh was to be there. At the last minute, a cub reporter found himself assigned to the event. He rushed to the scene, notebook in hand, and singled out the striking actress at the center of the crowd. Under his breath, he asked an onlooker who the actress was. He was told, "Vivien Leigh," whereupon he thrust himself through the crowd and questioned her loudly "Tell me, Miss Leigh, what part did you play in the film?" The story has it that Miss Leigh turned and walked out of the press conference.

As the Boy Scouts of America advise, be prepared.

The first preparation for an interview is to learn as much about your subject as you possibly can. The preceding chapter, "Know Your Subject," told you how to research a technical product. Now that you have background knowledge, it's time to prepare the questions you will ask.

PREPARING INTERVIEW QUESTIONS

Preparing a list of interview questions can help you in a number of ways. It lets the engineer know that you have given thought to the interview and that you value his or her time. The engineer is more willing to answer questions carefully knowing that the questions were carefully preplanned.

A written list also helps you keep control of the interview. Often, the interviewee will answer only the first few questions in order before unwittingly jumping ahead to a later topic, or going on a tangent that brings other questions to mind. Hugh Sherwood, in *The Journalistic Interview,* observes that "most interviews become unstructured as they progress, and a written list of questions gives the interviewer a structure he can return to whenever he feels it necessary."[2]

A questions list can make the job of writing easier. You can arrange your questions in the topical order you plan to use for your document.

It only follows that you'll ask the questions, and write the answers to them, in that order. Later, when you write your document, you can do so without having to jump back and forth through your notes. The information is already in the order you intended to use.

Write down your interview questions. Write more questions than you can ask in the time allotted. That way, you can be sure you'll take full advantage of the engineer's time. You won't have to cut the interview short because you ran out of questions (as I did in my first-ever interview).

Choose Questions That Work

> I keep six honest serving men
> (They taught me all I knew);
> Their names are What and Why and When
> And How and Where and Who.

Just about every English-speaking journalist has been taught this Rudyard Kipling verse. It summarizes the basic questions journalists must ask in almost every interview. These questions apply for technical writers, too.

To call these questions "serving men" is apt, because they serve you in any interview situation, even when you don't know much about a subject. If you've been unable to find out much background information, or if a product is totally new, you can still ask the engineer:

- *What* is the purpose of the product?
- *Why* would the customer want it?
- *When* will it be put on the market?
- *How* does it work?
- *Where* can I see a working prototype?
- *Who* is the customer?

Order Questions Logically

Arrange your questions in some logical order. As suggested earlier, you can arrange your questions to match the order of information in the document you're writing. Another arrangement you might choose is the order of information in an existing document, such as the product specification. Often the engineer you interview is the author of the product specification. In this case, use sticky-notes to keep track of questions on the specification, turn to the page with the note on it, and ask your question, perhaps quoting the passage you're not sure about.

A third way to arrange questions is to move from general questions to more specific ones. Journalists call this a "funnel" sequence: moving from more "open" (general) to more "closed" (specific) questions. When you move from general to specific questions, the engineer has a better chance of following your train of thought and perhaps providing additional information that you hadn't thought to ask for. You begin with your "What is the product's purpose?"; then proceed through more and more tightly defined questions, until your questions can have only one specific answer, for example, "What baud rates does the modem support?"

Group specific questions by subject. In other words, don't ask a question about a particular hose, then one about the release date for the product, and then another question about a hose. Move methodically from subject to subject. And move methodically within each subject. For example, move from general to more specific questions, or ask questions chronologically. This doesn't mean the interview will tightly follow your chosen order, nor should it. But by grouping your questions logically, you will keep track of them easily and think more clearly about the information you need.

You will most likely have to skip around during part of the interview, asking questions out of order. So number your written questions in advance. Then place a check mark next to each one as it's answered. That way, you can glance at your list and know which questions remain.

Phrase Questions Clearly

Make sure you phrase your questions clearly and in detail. That way, the engineer will understand each question and not have to ask you to repeat or explain it. This means including background information.

Let's take a real example. To research an index generator—a computer program that makes an index for a document—you need to find out how the user includes terms in the index.

Your first question might include some introductory comments: "This software generates an index for a document. But I don't quite understand how the user tells the software which terms to index. Would you explain how the user does this?" The introductory remark places the engineer right where you are. Your ensuing question lets the engineer know exactly the kind of information you want.

Basically, you are going to ask the engineer all the questions about the product that you haven't been able to answer elsewhere. Sometimes, you won't have answers for some of the simplest, most mundane questions and you'll be stuck asking the engineer. This brings up a critical point. Whether fresh out of school or very experienced, a technical writer always has to be willing to *risk looking stupid*. "He who asks is a fool for five minutes. He who does not is a fool forever." (Chinese proverb)

Sometimes the only way to find something out is to ask the expert. And no matter how long you're at it, there'll always be some term or concept you don't know, particularly in complex, rapidly evolving technologies. You will have to hear the incredulous "You mean you don't know what a supergizmokaffoble is?" and just hang tough. Take it on the chin, as the saying goes. If you fail to ask about the obvious, you might end up omitting it from your document.

▪ ▪

Checklist 8 – 1.
Preparing Interview Questions

The following checklist summarizes the tips discussed in this section. You can use it as a checklist when you prepare questions for an interview.

- Choose an order for your questions.
- Move from general to more specific questions.
- Group questions by subject.
- Phrase questions clearly.
- Ask specific questions, aimed at getting the kind of answer you need.
- Avoid yes/no questions.
- Number your questions.
- Don't be afraid to ask the obvious or to appear stupid.

▪ ▪

WHO IS THE INTERVIEWEE?

Before interviewing the engineer, find out as much as you can about him or her. Does the engineer have a reputation for cooperating with technical writers? Or is that person supposed to be an ogre? Ask your manager or other writers what their past experience with the engineer has been like. But don't let a bad reputation prejudice you. You might be the first writer to favorably impress that engineer with your courtesy and intelligence.

SETTING UP THE INTERVIEW

Phone the engineer to arrange an interview *after* you've organized your questions. That way, you're ready if the engineer says "Can you come over right now? I'm leaving for Argentina in an hour and won't be back for two months."

When you phone the engineer, introduce yourself and give your credentials for taking his or her time:

"This is Gary Meade. I'm a technical writer in the Software Documentation Group. My boss, Sheila Channing, suggested you might be the person to answer some of my questions about the Software Troubleshooting Guide I'm working on."

Next, tell the engineer the kind of information you need. Make sure to tell the engineer why you need him or her specifically, unless it's obvious. For example, if you need clarification about a product specification that the engineer authored, it should be obvious why you're calling.

If you need general product information, a busy engineer might try to refer you to someone else. If you've been told that he or she is the best authority on how the product works, mention it now. Make sure the engineer knows that he or she is a valuable resource for you.

Usually you will work with an engineer whom I'll call your "primary technical resource." This particular engineer will be responsible for the technical accuracy of your document. You will turn to this person for answers to your questions, as well as for leads to other authorities who can help you. If your primary technical resource cares about documentation, he or she may also pave the way by telling other engineers to expect your calls. He or she will be your key reviewer, when you send your written drafts out for feedback. You can find details about the key reviewers' roles in the chapter "The Review Process."

Sometimes your primary technical resource is a relatively inexperienced technical person who is willing to track down information for you. He or she may be more graciously received than an inexperienced writer by the upper engineering echelons of some companies.

SCHEDULING MORE INTERVIEWS

You will meet with your primary technical resource on an ongoing basis and sometimes work very closely with that person. You need to tell the person about how often you anticipate needing to meet, and how long each meeting should take. Engineers who've had experience working with writers will usually ask for this information. They need to know, so they can include the interview time in their schedule and let their manager know how their time is being spent.

Estimating how much of an engineer's time you will need is tricky. Every writer seems to need different amounts. And the amount of time will vary between projects. I strongly recommend never asking for more than an hour per interview. This is because you need to be very sharp during the interview. You need to listen intently and digest vast quantities of new, complex information. More than an hour of this is a burn-out. And the engineer gets burned out, too. Additionally, an engineer's schedule is tight. He or she may postpone meeting with you if you ask for more than an hour. Sometimes you may only be able to get 15 minutes.

You should press hard to find one dependable technical resource who is willing to meet with you for an hour at least twice a week, even if you don't feel you need that much time. Otherwise, gathering enough infor-mation later may become gruelling or impossible.

Start by meeting with your primary technical resource twice the first week. Then estimate how much time you'll regularly need, based on how many questions were answered in those two meetings and how much more information you'll need. You'll meet more frequently during the research part of a project than when it's in final editing. But make sure to establish an ongoing relationship with the engineer, in which he or she expects your requests for time.

TO TAPE OR NOT TO TAPE

To capture the responses to your interview questions, you'll either use a tape recorder or write detailed notes. Again, the similarities between technical writers and journalists come to mind. Both must decide whether to tape or take notes; both have personal preferences. There are a number of good reasons to use a tape recorder:

- When a topic is very technically complex, you can record concepts and statistics accurately.
- You can maintain eye contact with the interviewee.
- You can listen more fully. You know all the details are being captured by your tape recorder, so you can concentrate on understanding the bigger picture.
- You're free to think about tying in your next question, because you're not frantically taking notes.
- You're freer to reorganize your questions creatively, should the interview take an unexpected twist.
- You'll never have to ask the interviewee to stop talking while you finish writing something.

IF YOU TAPE

If you tape, remember to turn the tape recorder off during interruptions. For example, if the engineer answers the phone or someone drops by with a message, don't record the event. If you turn the recorder off after an interruption or social digression, remember to turn it back on.

One very good reason not to use a tape is the amount of time it takes to transcribe it. Taping is a two-step process: First you record; then you transcribe, or take notes from, your tape. Transcribing a tape almost always takes longer than the time the interview took, unless you are lucky enough to have a secretary who'll transcribe the tape for you. Taking notes is a one-step process: If you can take good, detailed notes quickly, you can work directly from them.

If you use tape, transcribe it within a day or two. Within that time span, you can usually remember thoughts you had during the interview that you might want to write down.

IF YOU TAKE NOTES

You may choose to take notes rather than use a tape recorder. You may not feel comfortable relying on a tape recorder or you may be a whiz at shorthand. Perhaps your topic is simple, and you can get most of what you need by taking notes and making a few sketches.

If you take notes, rather than tape, go over them immediately after the interview. You probably heard more details and nuances than you were able to write quickly. You can write them now, without the time pressure of the interview. Human memory is a fragile recording device. Most impressions and extra information will evaporate in a day or two, whereas they're still fresh after the interview. Also, if your rushed handwriting is hard to read, you'll be better able to decipher it immediately. You'll remember what you tried to write.

 ## THE INTERVIEW PROCESS

A tech writer friend of mine always begins an interview by scanning the engineer's office, then asking about the photo or trophy on his desk. If no photo or other knickknack is in sight, this writer focuses on the engineer's tie, coffee cup, whatever, and offers a compliment. When Bob first recommended this tactic, I thought he was a cad. But after a while, I came to see the wisdom in it. I don't recommend insincerity. But an opening remark that's somewhat personal can relax the atmosphere. In *Creative Interviewing, the Writer's Guide to Gathering Information by Asking Questions,* Ken Metzler advises journalists. "Use of small talk tends to identify the conversation as a human one rather than a mechanical one."[3]

For most interviews, keep your opening remarks fairly brief. Your time, and the engineer's, is valuable. You can make your ice-breaking overture while you're setting up your tape recorder or readying your pens and notebook. Then launch briskly into your first question.

ASKING FOR AN OVERVIEW

Asking for a product overview is a good way to start a first interview with an engineer. Getting a product overview from a technical person's point of view is particularly valuable if you are a new writer or the product is new. It also helps even if you are an experienced writer working on a known product, because it gets the interview off to a good start. The engineer talks for a while. You sit back, listen, and take notes or nod encouragingly. Not only do you learn about the product, but how the engineer thinks about the product. And more importantly, how the engineer *thinks*. Does he or she go on tangents and jump around a lot in explaining things? If so, you can be ready to take firmer control of the interview.

Does the engineer use a lot of jargon and unfamiliar terms? List them as you listen to the product overview. When it's ended, ask the engineer to define those terms for you. He or she will probably use less jargon in answering your subsequent questions.

WINNING THE ENGINEER'S RESPECT

You are always more fortunate to have an interested, supportive engineer working with you than a reticent, critical one. However, the outcome of the interview is largely up to you. It's up to you to come prepared; up to you to make a good impression; and up to you to win the engineer's respect. Winning respect is easier the more experience you have. After being a tech writer for a while, you will know how to organize questions, time your interview, and use the information well, even in a new subject area. Also, the more technical expertise you have in a subject area, the easier it is to win an engineer's respect. No question about that. However, you can enhance your working relationship with an engineer even if you know next to nothing about things technical.

First treat the engineer with respect, but don't be deferential. Being deferential puts you at a disadvantage. It's important not to let the engineer think he or she is doing you a favor.

Intelligence is an important quality that wins engineers' respect. Unfortunately, engineers will tend to judge your intelligence in terms of your technical expertise, even though the two are unrelated. If you seem to know something about the product and if you assimilate new information quickly, you will gain more respect. But be ready to admit you don't know something. By admitting ignorance, you show that you're willing to learn; to correct your ignorance. That is a sign of intelligence that should win respect from all but the dourest engineers.

When I first started as a tech writer, I had to admit ignorance a lot, and my admission didn't always go over too well. At the time, I wasn't making very much money and I knew the engineer was making lots, so I'd quip, "Hey, if I already knew what you know, I'd be making your salary!" That usually got a chuckle and made the information flow more easily. This, I suppose, illustrates that humor can sometimes win respect. At least it's worth a try.

Other weaponry that win respect are communication skills, especially active listening. These skills are the subject of the next section.

▮ ▮

Checklist 8 – 2.
Winning Respect

You can use the following checklist to review the guidelines presented in this section:

- Approach the engineer with respect, but don't be deferential.
- Organize your questions well.
- Research your subject as thoroughly as possible.
- Listen carefully to the engineer's answers.
- Admit when you don't know something.
- If all else fails, try humor.

▮ ▮

COMMUNICATING EFFECTIVELY

Sherwood recommends to journalists, "If you are a wise interviewer, you will go into the interview with the secret assumption that you will like the person you are to interview." Just as he or she must receive a good impression of you, you must come prepared to regard the interviewee in the best possible light.

If you feel immediate antipathy toward the engineer whose cooperation you must have, put that antipathy aside. If you can't find anything to like, regard the person as a valuable source of information. Focus on the information you need, rather than the source. Direct your attention toward gathering that information as inoffensively as possible.

Listening

Listening is the primary communication skill for all occasions. You can't get along with your friend, spouse, parent, child, or boss if you can't listen effectively. Listening is just as important to the interview process.

You and the engineer probably differ in the ways you think and express yourselves. You may have to rephrase your questions so that particular engineer can hear what you mean. And you will certainly employ "active listening" when he or she answers you. To actively listen, you repeat back what's been said to you in a slightly different way—the way you heard it. Here's an example of such a dialogue:

Engineer: For a point-to-point configuration, the asynchronous modes work better than the normal response mode.

Writer: You're saying that asynchronous communication is more efficient for point-to-point connections?

Engineer: Yes.

Writer: Why?

Engineer: Because there's no polling overhead required.

Writer: Then, polling takes time and makes the normal response mode less efficient?

Engineer: That is correct.

You get the idea. You bounce concepts back and forth, like a ball, to check your perceptions. But here's an important point. You're not just checking the accuracy of your understanding. You're also letting the engineer know that you are really listening—that you're a real human being sitting there taking in all this wisdom and knowledge. Communication is happening.

Expressing

The way you express yourself affects the interview. If the interviewee seems perplexed by your questions, try stating them differently. You might try analogies and metaphors. For example, you can describe a machine as an animal, consuming and defecating, or as a symphony, with different parts playing solo or accompanied by other parts. You might need to speak very literally with a particular person. Another person might understand a sketch better than your words. Try drawing a picture of how you think something works. Then ask the engineer to correct it or to draw a better one. Think creatively and express yourself flexibly in the interview.

Nonverbal Cues

An accepting tone of voice and facial expression go a long way in establishing rapport. If you make eye contact, nod encouragingly, and smile appropriately, you'll get more careful, complete responses than if you ask the same questions in a clipped monotone and bury yourself in note-taking.

Controlling the Interview

Interviews can easily get out of hand. You and an engineer have very different views of a product—and for good reasons. The engineer has spent loving hours in the bowels of a program or device or molecular model. This person sees it as a clever solution to a design problem, and you see it as an end-product someone needs to know how to use. The engineer may be willing to talk for hours about the product's technical details. But that's not what you need.

Here's how one technical publications manager describes the problem of engineers who are too close to their job to understand what the writer wants:

> They know the information in their heads and they just can't tell you about it. Try to get them to explain an overall concept of what this product does or give a twenty-five-words-or-less description about this project that you're working on, and they immediately zoom from that down into the seven-layer protocol and what the twenty-fifth bit in the second register does. This doesn't help the writer develop any sort of framework for the information.

What you need in an interview is the information that will help your reader, plus enough extra product knowledge to give depth to your writing. You will have to interrupt the engineer who launches into lengthy, detailed technical discourse. Courteously, but firmly, focus the discussion:

"Can I interrupt? I'm enjoying what you're saying, but I have many more questions that I need answers for. Could we get to some of them now?"

One technique I use with engineers who are too close to their work is to remind them who my audience is. I sometimes have to do this several times during the interview:

"The manual I'm writing is for the person using this security lock. They won't be aware of the process you're describing, but they need to know how to use the lock. What does that person see when the lock denies them access?"

On the other hand, some engineers have so much history with a product or technology that your interview with them turns unexpectedly into a private seminar. I have found myself listening to an expert who is so knowledgeable and gripping that I've shelved my questions for a future time, settled down, and listened.

Dealing with Reticent Engineers

The best thing you can do with a reticent input person is to ask for information in small chunks. Like being audited by the government. When they ask you for a tax audit, they tell you *exactly* what they want to know about. When you're going to a developer that you're having trouble getting information from, the day before your interview, give them a list of questions and say, "This is the information I'm digging for." And then go in there the next day and sit down and talk to them about it. The more prepared you are for what you really want, the more likely you are to be successful.

Thus, free-lance writer Linda Lininger offers a way to deal with reticent engineers. An engineer can be reticent in an interview for a number of reasons. He or she can be reticent for all the reasons that any human being might be: she might not be a particularly gregarious person, she might be having a rotten day, or she might have problems at home. I call this communication difficulty the human factor.

Two other communication difficulties are more endemic to the writer-engineer relationship: (1) The engineer has a superior attitude toward the writer, and (2) the engineer has a poor command of English. Unfortunately, every writer has to learn how to deal with these two communication problems, even though most engineers are articulate and supportive.

An engineer with a seemingly superior attitude may in fact be having difficulty answering your questions. Frequently, such a person will say "You don't need to tell the reader about that," when he or she is stumped about how to explain the answer to you or is concerned that the answer will prove complex and time-consuming. Tell the reticent engineer that you need to understand the product well enough to write about it, without always calling for help. Explain that the more you understand now, the less help you will need later. The prospect of your future independence should motivate the engineer to tell you everything you need! After you have explained why you want the information, rephrase your question.

I have good luck with a trick that seems as though it shouldn't work. When an engineer tells me "You don't need to know about that," I answer, "But, I'm just curious. Could you explain it to me?" And he or she always does. This approach works to open the flow of conversation because the engineer hears that I'm genuinely interested in what he or she is doing, that I'm not just doing a job. You will find tricks that work for you.

Breaking through the Language Barrier

Sometimes your primary technical resource speaks English as a second language and has not mastered conversational English. He or she has a strong technical background but rudimentary English skills. This problem is not serious if the engineer is willing to spend time and energy communicating with you. The language barrier becomes a problem only when the person is defensive about his or her lack of English skills or is not willing to overcome the difficulty of communicating complex technical information to you.

You need to be patient in this situation. Be willing to rephrase questions several times, draw pictures, try to use the product while the engineer watches, or whatever it takes. Assume that you are responsible for making the communication work.

Sometimes you'll do everything possible to communicate with an engineer whose English skills are minimal and you'll fail to get the information you need. In this case, you must find others who know the product and can communicate. Get them to answer your questions.

I recently reached a dead end with a non-English-speaking programmer and couldn't find another engineer who knew the product. I ended up talking to a writer in a different part of the company, who'd already documented the product for a different audience. He was very technically knowledgeable. He agreed to be my primary technical resource and the main reviewer for my document.

▪ ▪

Checklist 8 – 3.
Working with a Non-English-speaking Engineer

In summary, here are some guidelines for working with an engineer whose English skills are minimal:

- Phrase your questions in simple, clear English.
- Be patient—first assume the problem is with your communication skills and be prepared to rephrase a question several ways.
- Draw pictures, when possible and appropriate.
- Ask the engineer for the names of other technical people (including writers) who've worked with the product. Then go over your questions with them.
- Ask to use the product or watch the engineer use the product.

- Seek a translator who speaks the engineer's language and whose English skills are better.

■ ■

Double-check technical information you get from a non-English-speaking engineer. I wrote a document based on an interview with such a person and later found many of my understandings to be inaccurate. The language barrier can introduce more inaccuracies than you realize.

WRAPPING UP THE INTERVIEW

When either all your questions are answered or the amount of time you've asked for has come to an end, you will close the interview. Thank the engineer for his or her time. Mention—unless it's blatantly not true—that you got a lot of valuable information:

"This was an excellent interview. You gave me all the information I need for the first three chapters. Do you mind if I call you if another question or two comes to mind?"

Letting the interviewee know that you got a lot of information helps him or her feel that the time was well spent. That person will be more willing to meet with you again and will surely say "yes" to your request to make follow-up calls.

SUMMING IT UP

This chapter described the art of communicating with an engineer. It provided guidelines on preparing interview questions and making a good impression. It also explored the question of whether or not to tape record an interview. This chapter then described communication skills and other tips to help you maximize the value of your time with the engineer. Finally, this chapter gave tips on how to end the interview, so that the engineer graciously receives your next phone call or request for another interview.

The next chapter talks about communicating with the most important person of all—your reader.

■ ■

[1] Barry Farber, *Making People Talk* (New York: William Morrow and Company, 1987), p. 83.

[2] Hugh C. Sherwood, *The Journalistic Interview* (New York: Harper & Row, 1969), p. 36.

[3] Ken Metzler, *Creative Interviewing, the Writer's Guide to Gathering Information by Asking Questions* (Englewood Cliffs, N.J.: Prentice-Hall, 1977), p. 19.

9

KNOW YOUR
AUDIENCE

O ne day in May 1983, three of the four engines on an Eastern Airlines jumbo jet died in midflight. The passenger-filled plane abruptly dropped three miles before the pilot recovered and managed to land the plane. An investigation revealed the accident was not caused by faulty equipment, it was caused by the maintenance crew's failure to read the manual.

In this incident, reported by Jonathan Kozol in *Illiterate America*,[1] the crew had neglected to insert oil seals in the fuel line during a pre-flight check. In the investigation report, Eastern Airlines did not specify whether the maintenance crew ever opened the manual or opened it but failed to find or understand the instructions. Document usability experts might suggest this is a moot question—users stop opening a manual soon after they have difficulty finding or understanding information in it.

An electronics engineer I know, who is also a recreational pilot, says it's unfair to single out airplane mechanics:

> The consequences of their actions are very dramatic. How many computer technicians have failed to set some dip switch on a computer card and spent the rest of the day trying to make it work? Lots. I talk to them all day. But a similar error by an airplane mechanic might have more dramatic consequences.

Nobody reads manuals. Why the hell should an airplane mechanic read it? He's out in the rain, covered with grease. How can he read the manual?

A document is only effective if it reaches its intended audience. For your document to reach its readers, you first need to learn about them. This chapter describes the problem of reaching readers who can't read, and some real-world problems finding out about your audience. It goes on to describe ways to learn about your audience and tells how to apply what you've learned.

 ## THE PROBLEM OF ILLITERACY

Would knowing the manual's user have helped the writer prevent the Eastern Airlines accident described earlier? Studies have produced a profile of military airplane mechanics, who hopefully differ from the mechanics working on passenger jets. According to one report:

> The typical [aviation] mechanic is 24 years old, has a high school diploma, never enjoyed reading, and had a ninth-grade reading ability upon graduation from high school. Through nonuse, reading ability has deteriorated to almost the seventh-grade level. The typical individual does not use the traditional technical manual . . . for much of his work; he prefers to use proceduralized, task-specific maintenance repair cards. He gives no indication that he understands how or why the system works.[2]

A reasonable response to this discovery would be to produce extremely simple manuals with lots of illustrations, and that's what the military did. They created "New Look" Skill Performance Aids that depend heavily on pictures. Perhaps they even went too far. According to Kozol, some military documentation has become so cartoonlike and simplistic that one manual allowed five pages to explain how to open the hood on an army vehicle.

In the end, no amount of technical writing genius can compensate for a reader's illiteracy. Illustrations didn't help the Navy recruit who damaged $250,000 in delicate equipment, according to a report described by Kozol, because the recruit could not read.[3] He had attempted a repair by following only the illustrations. While the Navy recruit is an extreme example, your average reader might not be that far ahead of him in ability or willingness to read.

You cannot teach your audience to read or force them to open your manual if they're unwilling. But by knowing your audience's reading level, preferences, previous experience with similar products, and how they'll use your manual, you can direct your manual to their needs.

 THE REAL WORLD

Despite a long-standing awareness of the importance of knowing a document's audience, technical writing departments do not often look beneath the surface of what they already know or assume. They place readers in two very broad categories—technically naive and technically sophisticated. Common "wisdom" dictates that you define all terms for the technically naive, while for the sophisticated, you assume knowledge of terms and concepts that are "commonly used in the field."

Writers are left to make broad judgment calls on these issues. They are frequently urged on by harried technical experts who are quick to reassure writers "you don't need to tell the user that. He already knows," when the expert might not (1) know the answer, (2) be able to explain it easily, or (3) want to take the time to explain it.

The more you really know about the reader, the more you can know when to stand up to such pressure. At a recent meeting of the Society for Technical Communications (STC), Stephanie Rosenbaum, President of Tec-Ed, told a story about a writer producing a document for an engineering audience. Whenever the writer sought help defining a term or explaining a concept, the engineering resource for the project explained "Engineers already know about that. You don't need to put it in the manual." The writer decided to comply with all the engineer's advice about what fellow engineers needed in the document. Then, to test the document's fitness to its target audience, she sent a review draft across the country to an engineer in the field. The engineer sent back copious comments about the inadequacy of the document, its lack of definitions, explanations, and so on. The writer then showed these comments to her recalcitrant resource person. She never had difficulty getting information from him again.

This writer's ingenuity paid off. Most of the time, you'll be told rather tersely who the manual audience is, and that's all you'll get. The section "Getting Acquainted with Your Audience," later in this chapter, tells you some ways to get to know your readers better. But first, what specific characteristics of your readers do you need to know?

 CHARACTERISTICS OF YOUR AUDIENCE

The characteristics of your audience will guide you to the appropriate form and content for your document. For example, how would you provide an airplane mechanic who is "out in the rain, covered with grease," with appropriate documentation? How would your knowledge of his circumstances dictate form and content? One solution might be to give him a task card, laminated in heavy plastic, with short, clear instructions describing only those steps needed to get the job done. If you do not know the characteristics of the reader—in this case how he uses the product— you might choose an inappropriate form of communication.

Some characteristics to consider include:

- The reader's education level or reading ability.
- The reader's knowledge of the product technology.
- How the reader uses the product to perform a task.
- What the reader wants from a document.
- How many distinct audiences will use the document.

Does your document have two, or even three, distinctly different sets of readers—for example, a programmer, a system administrator, and a naive user? Are the customer and product user different people? Are the product user and document user different? These questions are not obvious, and writers don't often think to ask them.

 WRITING FOR A SINGLE AUDIENCE

If you have a say in planning the documentation effort for your department, strive to define a single audience for each manual in a product's documentation set. This set might include online help and video, and these too should have a single defined audience.

I was not so fortunate on one contract, where I was asked to write a user's guide for a data communication product. While the audience had not been clearly defined, the deadline loomed. Assuming I was writing for the program user, I began with a simple overview of what the product did, defined all terms in a glossary, and went on with step-by-step instructions telling the user how to use the program.

In the middle of doing this, my client requested that I tell the network administrator (audience #2) how to configure the software. I suggested a network administrator's guide would be the place for such information, but time and funds were not set aside for such a manual. I added a chapter for the network administrator.

Then the client wanted information included about how to configure a host computer at the other end of the communication line. The host system administrator became audience #3.

Next, an out-of-breath programmer whisked a software specification under my nose at the very eleventh hour. His specification would tell a programmer (audience #4) how to design a communication program that would work with the product. The spec was almost unreadable, containing undefined terms, passive voice sentences, and poorly formatted examples of code. According to the programmer, I only needed to "clean it up a little. The audience already knows about most of this stuff."

No, that was not the final blow to this manual. Someone in technical support insisted that I document his diagnostic program, which would help customer support engineers (audience #5) trace problems with data transmission.

The "user's guide" ended up with five distinct audiences, each with vastly differing needs, expectations, and levels of technical knowledge. My client was happy with the end result, which I considered a disaster.

Make every effort you can to define a single audience for your documentation project. If you're stuck with multiple audiences, at least give each its own chapter and clarify the audience up front. For example:

"This chapter addresses the system administrator. It tells you how to install and configure the software. If you want information on how to transfer files, refer to the chapter, 'Using the File Transfer Facility.' "

Once you've defined the general audience, you can get down to specifics about it. The rest of this chapter assumes you will have the time, money, and supportive management you need to learn about your readers and to implement documentation improvements based on your discoveries. In other words, it's based on an ideal—not necessarily the real world—and you might have to settle for less.

▌ GETTING ACQUAINTED WITH YOUR AUDIENCE

You can get to know your audience by talking with people, like customer support engineers, who know something about them or by talking to members of the audience. If you have the wherewithal, you can also test the usability of your preliminary documentation with members of your target audience.

If your situation doesn't allow for formal means, use any means you can to learn more about your readers. When consultant Daunna Minnich needed to write a workstation user guide, she was able to find representative users within her company:

> I was writing about what they now would call a workstation, and they had at Memorex everybody from engineers to data-entry clerks, which was ex-

actly who the audience was. It was everyone from people who knew nothing to people who could design the system.

Daunna astutely realized that she too was a member of her target audience and drew some valuable inferences from her own experience:

> Partly I used myself as a resource. This is the job—in the Memorex Communications Division—where I walked in for my interview thinking, 'Oh, this is the communications division, so they must do all the publicity stuff here. Isn't that great. Everybody will be writers and into publicity and stuff.' I had no idea we were talking about *tele*communications or *data* communications. I was totally stupid!
>
> I must have figured it out somewhere between going in for the initial interview and being hired, but I didn't know what that basic term [communications] meant.

By being honest about herself and having the self-confidence to share her observations with her boss, she was able to convince him to change the nature of their manuals.

> One of the things that I did in that book was to spend four pages defining some basic terms. "Here's a monitor, here's a keyboard," and back in 1980, not everybody knew what a monitor was if they were a data-entry clerk, whereas my boss said, "Everybody knows what a monitor is." Not if you're a data-entry clerk fresh out of high school you don't.

TALKING WITH CUSTOMER CONTACTS

While employees in your company might not accurately represent members of your target audience, you might still be able to gather information within your company. As part of their jobs, some employees in your company have direct contact with your audience. The most valuable of these are in customer support, training, and sometimes marketing.

The customer support engineer ("technician," "analyst," or "representative"), who provides either telephone or field support to customers with problems, is an invaluable source of information. He or she knows where users get stuck using the product, the kinds of technical errors that can cause the product to malfunction, and the product quirks the user has to learn about. By asking customer support people about the calls they receive, you can unearth inadequacies in the product's design and documentation. The support engineer usually knows the current manual and can make valuable suggestions about the kinds of additions that would save support effort. This person should also be on your list of reviewers.

Technical trainers are another good source of information about your reader. Trainers usually use the current manual, or pieces of it, in their hands-on classes for customers. Frequently trainers rewrite portions of the manual in response to problems that show up. For example, if the manual inaccurately describes a procedure used in a hands-on training class, instructors will rewrite that section. However, somehow they don't think to tell the manual's author about errors! My eyes were opened when I took a hands-on training class for customers and discovered the manual had been heavily edited and in places, completely rewritten, mostly to correct technical inaccuracies.

In the hands-on training class, I hoarded class materials and added the trainer to my list of reviewers. In addition, I talked with customers in the class to get a sense of their level of technical expertise and their problems using the product. I also watched how they used the manual with the product. Training classes are an excellent way to get to know your audience.

Marketing representatives often provide your initial information about your audience. With skilled questioning, you can glean a lot more from them than they first provide. As Stephanie Rosenbaum reports in a paper on document usability testing:

> Often, marketing groups develop a short audience description, full of buzz-words such as "novice computer users" or "power users", which they employ over and over in presentations and discussions.
>   For each buzzword, you need to interview for more information. . . . "How much experience can someone have and still be a 'novice'? What must someone know to be a 'power user'?"
>   By asking open-ended questions, you encourage the product manager or other respondent to think before answering. More important, you obtain specific operational information which will help you prepare rigorous user-group definitions.[4]

TALKING WITH YOUR AUDIENCE

If you can, obtain information directly from your readers about their education and skills, how they use the product, and what documentation they feel they need. Two ways to do this include a visit to a customer site and a reader survey.

Site Visit

A site visit lets you evaluate existing documentation, based on user feed-back, so you can improve your future efforts. If you can arrange a site visit, through your marketing or customer support departments, try to set up interviews beforehand with those customer employees who use the manual or online documentation. Be prepared to elicit the kinds of information that will help you improve your documentation. For example, ask users to perform a simple operation by following the previous documentation, and watch them. Do they thumb through the manual randomly? Do they look in the index or table of contents first? Are they so familiar with the guide that they turn immediately to the right place?

If they know exactly where to look in the manual, chances are they've regularly found it useful.

If they've had the manual awhile and are still unfamiliar with it, you might conclude the product is so easy to use that the manual is unnecessary, or the manual is so unattractive or difficult to use that the customer is happier fumbling blindly with the product, or some combination of product user friendliness and document user hostility. After all, if there's user friendliness, there must also be user hostility!

Questions to Ask Users

Augment your observations with questions. The following list provides questions you can ask the user to evaluate a manual. It assumes the user has access to either the previous version of the manual or a preliminary draft of one you are writing.

About Organization

- Can you find the information you need easily?
- If not, what kinds of information are hard to find?
- Is there a better way to organize the manual so that you can get to what you need?
- Is the index adequate?

About Completeness

- Did you need to ask someone for help to use the product?
- What information could the manual provide such that you would not have to ask for help?
- Is the manual too technical for you to understand the information you need?
- If so, what parts require clarification?
- Is the manual technical enough?
- What more technical information would you like to see in it?

About Visual Presentation

- Would more illustrations help you?
- What kinds of illustrations would you like to see?
- Do you think the manual looks attractive?
- How can we improve its appearance?

About Appropriateness

- Do you have additional suggestions for improving the documentation you use with this product?
- Is there an additional kind of documentation—for example, a computerized tutorial—that would help you use the product?

A Reader Survey

You can use a reader survey to learn about your audience either after they've used a version of the manual or before they've even seen the product. If you're conducting a survey of current manual users, use the same questions (listed in the preceding section) that you'd ask at a customer site to determine the reader's needs.

A reader survey can also reveal the characteristics of an audience you are about to address with new documentation. One such survey, described by Heather Keeler in an article in *Technical Communication,* sought to discover three user characteristics: "how my readers process information, their preference for style and content, and their familiarity with printed material. I wanted reader responses to help me determine whether I should explain details with pictures more than text, use formal or informal language, provide overview information or just details, and use marginal notes and headings to help readers skim text."[5]

Keeler designed a survey containing affective questions a social scientist might ask. For example, she asks the reader to respond to statements, like "When I read fiction, I hear characters speaking in my head," with a range of reactions from "strongly agree" through "strongly disagree." She included questions about job classification, age, and education.

Keeler found, among other things, that her readers "more effectively process information through pictures rather than words . . . They prefer simplicity over formality . . . Readers also want documentation to provide the whole picture, not just the details, and to include why something happens, not just how."[6]

Because Keeler surveyed an entire company, her respondents included audiences for a variety of documentation. By differentiating them, she was able to see how, say, manufacturing documentation needs more visual elements than are needed in documentation for a more print-oriented audience.

One of Keeler's final recommendations is "if in doubt, ask" the user.

USABILITY TESTING

"Usability testing is very simple," say Carol Bergfeld Mills and Kenneth L. Dye in an article in *Technical Communication:*

> You just find users who know nothing about a product and have them use it with the manual. By watching where they have problems using the product or where they need help, you can identify problems in the manual (and sometimes in the product).[7]

Some would argue with the simplicity of "just" finding users who know nothing about the product. Selecting users for such a test can be a very detailed process. Some testers first formulate a measurable definition of the manual's prospective user; then match the correct reading level, learning styles, and other qualities with a group of people who will take the test. Says Stephanie Rosenbaum:

> Audience analysis achieves rigor where an organized typology of users has developed over time. For computer documentation, these standard user typologies consist of categories of user traits . . . They are the *dimensions* within which to measure potential subjects; when quantified, they show what a typical representative of a group should be like.[8]

This is a far cry from walking down the hall to see if folks in accounting can figure out your new keyboard map!

Why a usability test? Why go to all this trouble when you can simply ask users about their education and experience, adjust your document reading level and vocabulary accordingly, have the document reviewed, and be done with it? Because, say cognitive psychologists, we bring to any task a complex agenda of past experiences and associations. While not simple to measure, these associations affect how we think and act. By observing a sampling of typical users, we can allow for the eccentric differences people bring to tasks and observe some of their less predictable responses.

Says Candace Soderston in an article in *Technical Communication:*

> Psycholinguists have studied the cognition process extensively and agree that syntax alone does not determine when inferences will or need be made. Each

of us brings to the reading task a background, a framework of knowledge
. . . If I bring to the reading situation a full background on the subject at
hand . . . I may not notice an ambiguity that a reader with less knowledge
on the subject will notice. However, a less informed reader, after searching
long-term memory and finding no existing point of reference, will have to
juggle several alternative interpretations and, after making one choice (perhaps not the correct one), generate the bridging inferences with which to
attach the information in the knowledge framework.[9]

The Test Lab

Usability testing of both products and documentation is now done by a
number of in-house and consulting laboratories, as well as universities.
The labs operate in a similar way. The American Institute for Research
runs a Usability Engineering Laboratory in Bedford, Massachusetts. According to a report in *BYTEweek:*

> [The lab is] deceptively simple, consisting of two rooms connected by a one-
> way mirror with intercom links. In the test room, computers can be set up
> to represent different types of working environments. The test subject, who's
> usually a "typical user" of the product user study, attempts to go through
> the usual product learning process, while the designers of the product and
> cognitive psychologists watch through the one-way glass. At the same time,
> the process is taped by three video cameras, and the actual keystrokes the
> user types are logged and time-coded.[10]

I've seen two such tapes, produced by Digital Equipment Corporation and Apple Computers. In both cases, users were either thoroughly
confused or enraged with the product's manuals. Besides leading to product
improvements, the taped sessions made excruciatingly clear the need for
improved documentation.

Soderston describes usability tests she performs in IBM's Human
Factors Laboratory in Kingston, New York:

> We present our subjects with the task for the day, our written material, and
> the system. We then leave the room so that our presence does not interfere
> with their processes. While they proceed through the task, verbalizing their
> thoughts, we watch from behind one-way glass. There is a microphone in
> the test room, so we can hear everything that the subject says. Besides seeing
> the subject through the glass, we also have a telescreen monitor aimed at the
> display screen of the system, so that we can see whatever the subject types
> on the screen.[11]

The Results

What are the elements usability testers measure? According to Mills and Dye, "Three main classes of information are collected in usability tests: logs, objective measures, and subjective measures." Logs record testers' observations, users' keystrokes, and "verbal protocols" (thinking aloud). Objective measures include things like the number of errors made and the amount of time it takes to perform a task. Subjective measures include the users feelings and opinions about the product, which testers will collect through interviews and questionnaires.[12]

Conclusions drawn from these data depend on the kind of documentation and the nature of the tasks the user performs. You wouldn't need to be a trained social scientist, though, to understand the grosser problems expressed by test subjects in the sessions I observed. However, if you really want to do it right, you can hire a consulting lab to test your documentation. The cost? According to a 1989 article in *BYTEweek:*

> A full-scale test, which involves 10 to 20 sessions and several weeks, costs approximately $25,000. Even a minimum test, with one or two people, costs $10,000. Potential savings on support can, however, pay back many times that amount.[13]

The cost today would be even higher.

FOLLOW-UP

No amount of research will help you if you don't use what you've discovered to improve your documentation. You may think this goes without saying, but I've found it's not as obvious as it seems. At one company where users were disgruntled about documentation, I was sent to find out why. "Design a survey," my boss suggested. I designed one, with questions very similar to those you'll find earlier in this chapter in the section titled "Site Visit." I submitted it to user groups both in the United States and in Europe. Customers were obviously pleased to have their opinions solicited, because a large percentage returned my questionnaire. The results were easy to analyze, because the majority of users requested the same improvements: More illustrations, a better index, and more technical information on a more sophisticated level. They didn't care about the attractiveness of the manual's cover. They just wanted the facts.

I presented the results at a department meeting and made concrete suggestions about how to give the users what they'd requested. Everyone was in agreement. But it never happened. The survey, it turned out, had been a gesture of diplomacy. Management did not intend to implement changes. Seven years later, that company's customers still complain about documentation.

The lesson—enlist the support you'll need from management and fellow workers before you begin to investigate your audience. Prepare a concrete plan to give your readers what they need.

What are some of the elements you'll modify based on your findings about your audience? Some obvious ones already mentioned include level of technical complexity, amount of detail, number of illustrations, and formality of language.

When Keeler found a significant minority of her test subjects were "nonreaders"—that is, they read fewer than five newspapers, books, or magazines per month—she recommended using "lots of headings, marginal notes, illustrations, and bulleted lists."[14] These recommended elements are standard tools tech writers use to break up the printed page. More about that in the chapter "Planning for Visual Impact."

Other document features based on audience might be unique to your user and might require you to find a more creative solution. The example of the airplane mechanic in the mud is a good one, and the plastic-laminated task card is, I think, a good solution. Others, from an article in *In House Graphics,* further illustrate this point:

> If customers want to write notes on the document you're designing, give them uncoated paper and plenty of white space. Depending on how they write them, a ruled section that looks like notebook paper may be appropriate.
>
> If you're doing a publication for people who want to flip quickly to the specific sections, you'd better put in a good table of contents . . . Consider color-tabbing chapters to correspond to the table of contents.[15]

SUMMING IT UP

This chapter described ways to find out who your audience is and what they need from your manual to do their jobs. It recommended some ways you can use this information to improve document usability. The next chapter begins a new section, which describes the birth of a document and its passage through writing, editing, review, and production. It provides guidance for planning both paper and online documentation.

■ ■

[1] Jonathan Kozol, *Illiterate America* (New York: Anchor Press/Doubleday, 1985), p. 20.

[2] Thomas M. Duffy, et al., "An Analysis of the Process of Developing Military Technical Manuals," *Technical Communication,* Journal of the Society for Technical Communication, Second Quarter 1987, p. 70.

[3] Kozol, p. 18.

[4] Stephanie Rosenbaum, "Selecting Appropriate Subjects for Documentation Usability Testing," *Work with Computers: Organizational, Management, Stress and Health Aspects,* Proceedings of the Third International Conference on Human-Computer Interaction, Boston, Massachusetts, September 18–22, 1989 (Amsterdam, Netherlands: Elsevier Science Publishers B. V., 1989), p. 621.

[5] Heather Keeler, "A Writer's Readers: Who Are They and What Do They Want?" *Technical Communication,* Journal of the Society for Technical Communication, First Quarter 1989, p. 9.

[6] Keeler, pp. 10–11.

[7] Carol Bergfeld Mills and Kenneth L. Dye, "Usability Testing: User Reviews," *Technical Communication,* Journal of the Society for Technical Communication, Fourth Quarter 1985, p. 40.

[8] Rosenbaum, p. 622.

[9] Candace Soderston, "The Usability Edit: A New Level," *Technical Communication,* Journal of the Society for Technical Communication, First Quarter 1985, p. 18.

[10] "Software Usability Testing: It's Time Has Come," *BYTEweek,* February 20, 1989, p. 3.

[11] Soderston, p. 18.

[12] Mills and Dye, pp. 42–43.

[13] "Software Usability Testing," p. 3.

[14] Keeler, p. 11.

[15] "K.I.S.S. & T.Y.M. (keep it simple, subscribe, and telegraph your message)," *In House Graphics,* p. 6.

PART THREE

A DOCUMENT IS BORN

10

....

ORGANIZING A
WRITING PROJECT,
ON PAPER AND
ONLINE

This chapter begins the odyssey of producing a document. Along the way, you'll travel through the mental realm of organizing thoughts on paper; the visual world of graphs, tables, images, and page design; the businesslike trail of schedule milestones; the creative adventure of writing; the editorial forests of rules and semicolons; the diplomatic forays of the review process; and the final road to your destination—producing a camera-ready book.

This chapter describes planning a writing project and organizing its contents, both on paper and online. Paper documentation is still the dominant medium in technical communication. Thus, the chapters following this one proceed through the steps of producing a paper manual. But first, you need to know something about online documentation.

Online documentation is text and graphics displayed on a computer screen. Its purpose is to teach, guide, or inform the person using the computer. Most often, online documentation accompanies computer software products like games, word-processing programs, graphics software, medical records software, manufacturing inventory software, and computer operating systems—to name only a few. Common examples of online documentation include help messages and reference documentation, tutorials, error messages, and technical reports. Some of these can be visually exciting, and can include animation and sound.

Many predict on-line documentation will be the dominant medium in the future. Therefore, this chapter covers general principles of organization for online, as well as more detailed principles of paper document organization. It also looks briefly at the future of technical documentation in video and beyond.

THE REAL WORLD

Before looking at the ideal, let's look at how you can expect to organize a document on your first technical writing job. At least at the start, your document's style and organization probably will be dictated by your writing department. Even more likely, you'll be given a book to update or revise and won't have the opportunity to consider whether it's organized for maximum usefulness.

As you participate more in department decisions, you'll have a chance to suggest changes that will improve the organization of documents as well as changes to an entire product library. These may be your biggest contributions to document quality.

Time is an obstacle you'll confront in planning a document. The up-front effort of organizing your material into a purposive whole is not always valued or understood by nonwriting managers and other members of a product team. After all, writers just sit down and write, don't they? If you ask for a significant amount of time in the planning stage, you might be met with suspicion, unless you've already proven yourself to be a "fast" writer. (A fast writer is one who meets deadlines without obvious signs of panic.)

Despite real-world obstacles, careful preparation is worth fighting for. Even more than the ability to write tidy prose, your ability to organize technical information will endear you to your readers.

THE IMPORTANCE OF ORGANIZATION

Why is organization so important? A few reasons, out of many, follow:

- Organization is the key to comprehension and document usability. It directs the reader to your intended communication goal.

- A solid structure, expressed in a written outline, greatly speeds the writing process. You are no longer confronted with a blank sheet of paper. Your structure is in place, waiting to be filled in.
- An outline can be reviewed and revised. Reviewers can suggest changes to the document's organization early, eliminating the need for drastic changes after the complete manuscript is reviewed.
- You can eliminate redundancy between manuals written by different writers within the group, because each is aware of what the others' documents will cover.

THE DOCUMENT PLAN

The document plan (or "doc plan") is the map you'll follow to produce a document. Sometimes it's called a document design, a blueprint, or by some other name, and its characteristics vary among organizations. A doc plan usually contains the following elements:

- the document title
- a statement of purpose
- a statement of audience
- a brief product description, including version number or other identifying product-release information
- a list of reviewers and their departments
- a list of other documents that support the same product
- a detailed outline, including positions of figures, graphs, and tables
- a schedule showing milestones (described in detail later in this chapter)

Appendix C, at the end of this book, contains an example of a detailed document plan.

SCHEDULING

How long should the planning process take? "The drafting of blueprints normally takes from 25 to 30% of the entire writing time," says R. John Brockmann in *Writing Better Computer User Documentation*.

This may sound like a lot, but probably a great deal of what you now consider writing time is really planning time. And, even if this is a lot of time, the time spent developing writing blueprints is much more cost effective than unplanned time spent rewriting during the 'editing' or 'reviewing' stages that try to make up for planning mistakes or organization problems. The hidden cost of unexpected problems can be greatly diminished by planning.[1]

Brockmann's blueprints are very like the document plan described in the previous section. Brockmann's time estimate for planning is a generous one, but you can certainly use his arguments, and other reasons listed here, to justify more time in your schedule for this vital activity.

SCHEDULING CONSTRAINTS

Document scheduling is strongly affected by the schedule for producing the product. Often product deadlines inflexibly dictate document deadlines. This situation is particularly prevalent in organizations where the publications department is not highly valued and the publications manager doesn't have the clout or charisma to communicate writers' scheduling needs. Artificial scheduling constraints can force you to write a final draft about an unfinished product, so the manual will be printed by the product release date. You can imagine how such constraints produce inaccuracies—and angry customers.

Rules for scheduling a document are difficult to formulate, because writers require vastly different amounts of time during different documentation stages. One writer might research painstakingly and write at the speed of a comet. Another might be a very slow writer, but will polish style and check facts during the process, producing a much more finished draft than the comet writer.

SCHEDULING MILESTONES

When you write a schedule, you'll include dates for the following document "milestones":

- research phase complete
- document plan distributed to reviewers
- review comments due
- first review draft distributed
- review comments due
- second review draft distributed
- review comments due (sometimes a third review is required)
- final editing, table of contents, and index complete
- camera-ready text available
- pasteup by artist (if one is involved)
- camera-ready copy to the printer
- printed, bound manual available to customers

These stages will vary, depending on your production tools and responsibilities. For example, you or an artist might insert graphics into the same computer file as the text, eliminating the pasteup process.

(An example of a document schedule is included in the document plan in Appendix C.)

PRINCIPLES OF DOCUMENT ORGANIZATION ON PAPER AND ONLINE

The way you structure information directs your audience's thought process or actions toward a specific goal, be it to understand the results of a study, to install a device, or to manipulate a computer program. If you place information on the page as if you were writing a grocery list, with no regard to the relationships between ideas, the reader has no signposts to guide thought or actions and must organize—mentally rewrite—the information before using it.

Organization sets up relationships among the kinds of information you'll communicate and provides clear paths from one to the next.

SIMILARITIES BETWEEN PAPER AND ONLINE DOCUMENTATION

Both paper and online documentation share two organizational principles:

- Recognizable structure
- Task orientation (versus product orientation)

The first principle—to give the reader a clearly recognizable path through the material—is accomplished differently for paper and online documentation. For paper, you reveal the document's structure both through logical writing, which includes transitional words and phrases, and through rhetorical signposts, like headings, introductions, and summaries. Reference aids, like a table of contents, further help readers find their way.

Online, you provide navigational aids, like pictures of keys labeled FORWARD, BACKWARD, and QUIT, which allow the user to move about, and you maintain consistency in the ways you "link" information. (*Linking* is a term used to describe a connection between computer-displayed areas of information.)

The second principle has to do with the fact that most people use technical documents to help them perform a task—to do—rather than to learn something. Thus, most technical documents are task-oriented, as opposed to product-oriented. They are (or should be) organized to most efficiently guide the user to do a job.

Studies have revealed that of all the reading people do on jobs, only 15% is reading to learn—that is, to retain information. The rest is reading to do.[2]

Writing for the doer is called task orientation. Rather than writing about a product, you write about actions in the order in which the reader will perform them, or you provide reference information in the order in which it can most quickly be retrieved. When the reader completes a task, he or she puts the manual away—or "exits" the online instructions—and can even forget the documented information.

Research has shown that, while task-oriented documents take 42% more time to create than product-oriented documents, they increase user productivity by 41%. Also, users are happier—79% of those who were asked to compare the two preferred the task-oriented documents.[3]

How does task orientation dictate document organization? Redish, Battison, and Gold succinctly summarized some guidelines in their essay, "Making Information Accessible to Readers":

> Put yourself in the place of the reader, and ask the questions a reader would be likely to ask. Then order the questions logically (. . . by importance, timing, or location in a sequence of steps in a procedure.) Do a task analysis of the procedure you are describing. If possible, do the task. Write down the steps of the task in logical order.[4]

Document organization is described in detail under "Organizing a Manual," later in this chapter.

DIFFERENCES BETWEEN PAPER AND ONLINE DOCUMENTATION

While the need for recognizable organization and task orientation guides both written and online projects, these media differ in some important ways. Once you know their characteristics, you can choose the one most appropriate to your communication goal, assuming you have access to the necessary tools.

The following sections describe the differences between paper and online documentation. Table 10-1 summarizes these differences.

For more detailed discussions of the differences between paper and online documentation, refer to R. John Brockmann's *Writing Better Computer User Documentation*[5] and William K. Horton's *Designing and Writing Online Documentation*.[6]

Characteristics of a Book

"Books take their place according to their specific gravity as surely as potatoes in a tub." Emerson (*Journals*, 1832)

A book is a substantial object. You can touch it, heft it, and even throw it, should the need arise.

If you need to find information in a book, you remove it from a shelf, turn to a table of contents, flip the pages until you come to the right spot, read, then close the book, and reshelve it. A lot of physical steps are involved.

If a book is very thick, you might feel reluctant to open it. Once you do, you'll passively absorb what you find, proceeding in a fixed, linear direction. While you can move back and forth through its pages, the book's character is essentially static and linear. Nonetheless, you always know where you are in it.

A book takes considerable time to produce and usually requires an outside vendor to manufacture it.

Despite its fixed quality, a book can contain lots of useful information. It allows for long discussions and complex, detailed information. It's familiar. Everyone who can read recognizes a table of contents. A book is also portable. You can take it on trips or use it outdoors.

According to William Horton, documents that fare better on paper than online include (among others) those requiring lengthy, detailed readings; those needed away from the computer; and those for non-computer users.[7] These documents benefit from the linear, settle-down-and-read qualities of a book, a book's relative portability, and its familiarity to the technically uninitiated, respectively.

Characteristics of Online Documentation

Online documentation is intangible and dynamic. By merely typing a key, or using a pointing device called a mouse, you can select a topic and obtain virtually instantaneous answers to questions.

Your relationship with computerized documentation can be interactive, like a conversation. For example, you ask a question; the software provides a general answer, then displays the message:

Do you want to know more?

Good online documentation lets you select levels of detail and move

Table 10-1. Differences between Paper and Online Documentation	
Paper	*Online*
Physically tangible and static	Flexible and dynamic
Passive	Potentially interactive
Difficult and time-consuming to update	Easier to update
Provides slow information access	Provides instant information access
Users can find place within whole	Users can get lost
More legible	Harder on eyes; requires more white space
Better for presenting complex and detailed information; discussion can take up as many pages as necessary	Small screen area necessitates short, self-contained information chunks
Portable; can be taken anywhere	Only as portable as the computer
More appropriate for noncomputer products and computerphobic users	Requires some familiarity with computers

between different kinds of information very quickly and easily. Bad online documentation leads you into corners from which you can't escape. Unlike a book, which has a beginning, middle, and end, online documentation has no visible dimensions.

Online documentation requires a computer. This limitation makes it impractical for most portable applications (though laptop computers are rapidly changing this). You won't use online documentation to help you fix the lawn mower in your driveway; you'll probably use a manual.

Online documentation is restricted by the physical characteristics of the computer—a small screen and letters that are less clear than print. Screens require lots of *white space*—blank areas in which no text or graphics appears—to be legible. This requirement limits the amount and type of information that online documentation can effectively present. Small, self-contained "chunks" are best.

Additionally, to use online documentation effectively, you must be familiar with computers. While many schools teach children about computers at ever earlier ages, computers are still not as familiar as books.

Computerized documentation is faster to produce than a book. A diskette doesn't need to go to the typesetter or to be printed and bound, and it takes up far less space than a book.

William Horton argues that certain kinds of documents gain value from being online. Some he mentions include documents containing critical information (like medical records), documents that are changed frequently (like price lists), documents containing intricately cross-referenced information, and documents that are part of the product.[8] All these documents benefit from being updated more quickly and accessed more easily than they would be on paper.

 ## ORGANIZING A MANUAL

Following the principles of recognizable structure and task orientation, what are some ways to organize a manual? The following types of organization are familiar to most readers, and therefore fulfill the first principle—recognizable structure:

Chronological	Events are described in the order in which they occur.
By Need	Frequently needed information comes before information the user needs less frequently.
By Difficulty	Information builds from simpler to more complex or difficult concepts.
Question-and-Answer	Information anticipates the user's questions and provides answers.
Alphabetical	Information is tied to terms that can be alphabetized.

The book you are now reading is chronologically organized to match the steps in a beginning technical writer's career. It's meant to be read through first; then used for reference later. Chronological order works best for books that lead readers through a sequence of steps and for books that will be read from start to finish.

Ordering information by need works well for a manual where each chapter contains the same kind of information (e.g., procedures *or* reference material). Otherwise, this order can defy the user's expectations. For example, most experienced readers of technical documents expect tutorials to appear before reference information, even though reference information is needed more frequently than a tutorial. If you place reference material first and tutorial last, users will find the organization confusing. No matter how logical a structure seems, it won't work unless the user readily recognizes it.

In general, you'll need to provide better reference aids in a manual organized by need, because its structure is less readily recognizable to readers than some other structures. Later in this chapter, you'll learn how to use a "road map" to clarify this type of document organization.

Information organized by difficulty is common for a tutorial, or any teaching text, which begins with what the user knows and proceeds to teach new concepts based on familiar ones.

Troubleshooting material follows question-and-answer order. The writing sample (described in the chapter "Breaking In") that contained questions and answers based on customer service calls is an example of this order. Questions and answers need to be ordered further, however, either by need (most frequently asked questions first) or by difficulty (easiest questions first).

Common examples of alphabetical material include glossaries and bibliographies. Reference material describing computer commands is often ordered alphabetically.

OUTLINE FORM

You'll organize your document by writing an outline. Some writers skip this step. Don't. An outline is your document structure made manifest. It is the framework upon which everything else is built.

An outline can be tedious to write. Like the framework of a building, it is not as beautiful or creative as the final edifice. (The outline for this book sent me to the coffee pot more times that I will admit.) But once in place, an outline allows you to create the headings, transitions, and cross references that will make your text flow.

During whatever happens between high school English and the present, most of us forget how to write an outline, so let's review. Figure 10-1 illustrates basic outline form.

For nontechnical material, your outline can guide your discussion without each level corresponding to a heading. In contrast, for technical material, each move between topics and subtopics is always signaled by a heading.

Figure 10-1.   Outline Form

I.   First level
II.   Another first level
  A. Second level
  B. Another second level
    1. Third level
    2. Another third level
      a. Fourth level
      b. Another fourth level
  C. Back to second level

A few rules accompany this essential tool:

- You can move "in" only one level at a time; you can move "out" as many levels as you want. For example, Figure 10-1 shows a second level (C) after a fourth (b).
- You must include at least two headings per level before you move back out. In other words, if you have one second-level topic, you must include another one before you move back out to the next main topic.
- Each heading must be followed by text, including the chapter title. No heading should lie next to another heading like bread in an empty sandwich.

That's all there is to an outline. Now let's fill it in.

HEADINGS

Headings are rhetorical signposts that tell the reader where you're going. As such, they should clearly express the contents of the section. Janice Redish et al. report:

> Research on how people read and understand has shown the obvious: headings help readers see the organization and understand the text. Research has also shown something that is not so obvious: vague or general headings can be more misleading than no headings at all.[9]

The level of a heading indicates whether what follows is a main topic or a subtopic. You càn indicate head levels either by the milspec numbered format, still quite common even in nonmilitary documentation, or by typographical emphasis. Figure 10-2 illustrates the milspec numbering system.

Typographical emphasis uses font size, capitalization, boldfacing, and underlining to visually increase or decrease the importance of a heading.

Figure 10-2. Milspec Numbering
1. First level
1.1 Second level
1.2 Another second level
1.2.1 Third level
1.2.2 Another third level
1.2.2.1 Fourth level
1.2.2.2 Another fourth level
1.3 Back to second level

Thus, a chapter heading might appear in all capital letters and 14-point type, followed by a level-one heading in 12-point type, and so on. The headings in this book use typographical emphasis to differentiate heading levels.

REFERENCE AIDS

Reference aids help the reader identify the structure of your manual. These include the table of contents, the index, dividers with tabs, "icons," and introductory road maps summarizing the contents of a manual.

Nothing prevents user-confusion better than a good index. All manuals should have one. You'll learn the principles of indexing in the chapter "The Production Process," later in this book.

Dividers with colored tabs are an inviting way to give the reader access to information quickly. They're particularly helpful in a large reference manual with distinct sections or a binder containing multiple manuals about the same product. Each tab should bear the name of the section or manual.

Icons are small symbols that alert a reader to a type of information. For example, warnings about electrical hazards are sometimes marked by a symbol in the margin of the text. You can use this technique to indicate other kinds of information as well.

An introductory road map is a fine way to lead a user to needed information. A road map is a table or diagram that summarizes information and indicates where to find it. Figure 10-3 shows a chronological road map I designed for a manual.

COMPONENTS OF TECHNICAL BOOKS

This section briefly summarizes the most common components of a technical book and tells you about the kinds of information you'll need to include in them.

Figure 10-3. Example of a Road Map: "QUICKSTART"	
To do this:	*Refer to this section of the manual:*
1. Set up your directories	"Before You Index" in *Chapter 3 Creating an Index*
2. Install the software	*Chapter 4 Installing AutoINDEX*
3. Collect text from drawing files	"Word Index" in *Chapter 3 Creating an Index*
4. Create the index	"Build Index" in *Chapter 3 Creating an Index*
5. Find drawing files	*Chapter 2 Using AutoINDEX*

Front Matter

The front matter includes the title page, copyright notice, trademarks, table of contents, figures and tables lists, and sometimes a preface. Its pages are numbered in small Roman numerals.

Check to make sure the copyright notice shows the year the book will be available and is not outdated. Make sure the trademarks for all products named in the book are listed correctly in the trademarks section. Sometimes a legal department will help you with this.

A preface usually includes a brief description of the book's organization. It can include a road map, like the example just given. It might also include a summary of conventions, described next. Be aware that readers hardly ever read the preface, so do not place essential information there.

Summary of Conventions

Often a technical manual includes a table showing how certain typographical conventions are used. For example, a computer manual might include a table, like the one in Figure 10-4, showing how keystrokes will appear.

Figure 10-4. Keystroke Conventions	
[ESC]	Escape
[CTRL]	Control
[F1]–[Fx]	Function keys F1 through Fx

You might also provide conventions tables for icons and their meanings, warning symbols, and any other visual cues that are not inherently obvious to the reader.

Overview

An overview, or introduction, gives readers a big picture of the product's organization. It summarizes the product's uses or features and places them within the whole. Like a preface, an overview is often not read. However, it's very valuable to new product users who are not familiar with similar products. And it can compensate for design flaws in the product by pointing out the unexpected. If a reader becomes perplexed about where to find information in the manual, he or she might turn to the overview to get a clearer sense of how the product works before proceeding.

Glossary

A glossary is an alphabetized list of technical terms and their definitions. It is an excellent addition to any manual. It helps new users and old alike. Often even engineers are unaware of certain terms in their field.

A glossary also gives you an excuse to pin down technical experts about their use of terminology.

Appendixes

An appendix is information that is "appended" to the book. It is not an afterthought, although it's occasionally misused as a repository of late-arriving information. Appendixes follow chapters and are assigned letters rather than numbers—Appendix A, Appendix B, and so on.

Any information that would be useful to readers but is not critical to your main communication goal should go into an appendix. Appendixes can include long tables, like conversions between different keystroke conventions, temperature (e.g., Celsius to Fahrenheit), and the like, as well as long lists.

The appendixes at the end of this book are examples of the kinds of information to include. Two are lists and one is a lengthy example. These appendixes provide information that is useful to some readers but that would interrupt the organizational flow if included in the main text.

Bibliography

A bibliography lists books on the same or related subjects. While many nonfiction books include a bibliography at the back, a technical manual will often list related titles at the front.

Index
This critical reference aid comes after all appendixes and the bibliography at the back of the manual. It's discussed in detail later in the chapter "The Production Process."

 ORGANIZING ONLINE DOCUMENTATION

As stated earlier, the main principles for online organization are the same as for books—recognizable structure and task orientation. How do these principles guide the organization of online documentation?

Just as paper documents can be ordered in obvious ways, such as chronologically and by need, online documentation follows certain structures that are recognizable. For online documentation, structure consists of the ways a user moves from one displayable unit of information to another. The path between two displayable units is called a *link*. The pattern or arrangement of links forms the structure of the online documentation, just as an outline is the framework upon which a paper document is built.

For example, in an online library catalog, a user can find information about a library book by typing the author's name into the computer.

However, just typing the name probably won't provide everything the user needs. Entering the name into the computer will locate a list of books by the author. To find information about a specific book, the user must search further. Here are some "online searches" a user might perform regarding the book:

- The user will need to find the specific title among several the author has written.
- The user will then want to know if the book is available or checked out.
- If the book is checked out, is there another library in the area that has the book?

Each of these kinds of information is linked to the author's name but appears on a separate screen. The screens can be connected in several different ways, either to one screen from which they all branch or to each other in one of a variety of patterns.

ONLINE STRUCTURES

For the organization to be understandable to the user, it must follow a consistent, recognizable structure. The example of the online library catalog follows what William Horton calls a grid structure.[10] You can type in an author, title, or topic; or choose to browse alphabetically through the library catalog. Any of these options leads you to a list of titles. When you select a title, the computer displays information about how many copies are in the library, their catalog number, and whether they are available or checked out. This structure is somewhat like a table— you look at a column heading, then move down the row of information for that column.

Another common structure is hierarchical—a displayed main list (or "menu") allows you to choose subtopics, each corresponding to its own screen of information. Some subtopics can be submenus branching to more detailed information or to yet another level of submenus. The main menu is very much like a table of contents from which you select a topic. The way the subtopics branch are very like the outline form of a paper document, in which a main topic is followed by subtopics.

These and other structures are clearly described in Horton's *Designing and Writing Online Documentation*.

ELEMENTS OF ONLINE DOCUMENTATION

Online documentation consists of displayed units of information and the links among them. Displayed units can combine small amounts of text with visual elements like boxed areas, shading and color, pictures of keys, and sometimes drawings. Positioning of these elements on the screen is very important.

On a computer screen, type can blink or appear in "inverse video"— light letters on a dark background. In the hands of the inept, a screen can look like a three-ring circus.

One way to clarify the organization of online documentation is to use these visual elements with controlled consistency. For example, if the user is to select a command from a boxed menu on one screen, you should not ask the user to type commands later in the same documentation. Commands throughout the documentation should be issued in the same way, either by selecting them from similar looking menus located in the same area of the screen or by typing them from the keyboard. Whatever way you choose, stay with it.

Consistency allows the user to feel familiar with the structure and to concentrate on the task he or she is trying to accomplish rather than the means to get there.

Similarly, visual elements should be stylistically consistent. If boxed text appears on a blue background on one screen, it should have a blue background throughout. Blinking text should always mean the same thing.

For example, a blinking message is sometimes used to signal the arrival of online mail from another user. You would not notice that your mail has arrived if the screen is dotted with blinking elements like a Christmas tree.

These are brief examples of a very complex medium.

NAVIGATION AIDS

Ways of finding screens, like the table of contents in paper documents, are the navigational aids directing the user. One way mentioned earlier is to provide a main menu that, like a table of contents, lists subtopics you can select and display.

Another way is to show a set of keys at the bottom or top of the screen with labels indicating what will happen if you press them. Some online documentation shows keys on every screen which allow the user to move FORWARD, BACK, or to QUIT the program.

Two additional navigational aids include *bookmarking* and *context sensitivity*. Bookmarkings lets the user mark a place in an online document, so that he or she can leave the document and return later to the same place. The familiar analogy of a bookmark is quite apt.

Context sensitivity is often used in online help documentation. When the user requests help, by pressing a help key or issuing a command in some other way, the software "knows" what action the user is trying to perform and provides help information specifically geared to performing it. Context-sensitive help is tricky to write because you have to outguess the user. As a user, I've been frequently annoyed with context-sensitive help messages that incorrectly guessed what it was I wanted to know and didn't provide a way for me to get the answer to my real question.

When you design online documentation, you'll either use an online authoring system to create it or work closely with a programmer who will write the software that links and displays your documentation.

▌ VIDEO AND BEYOND—A LOOK AT THE FUTURE

I'm not a real computer nerd, but even I could not resist an animated moose that would appear in the upper left corner of my Macintosh screen and say—yes, say—"How come we never go out anymore?" or "You are getting sleepier and sleepier." He would say things like this whenever there was a lull in my typing. This software allowed me to adjust the moose's tone of voice and the speed with which he said things. (Eventually, I jilted the moose. As I turned him off, he said "I wouldn't do that if I were you." I await his revenge.) The reason I'm telling you this is so you'll see how even at this beginning stage in the history of online documentation, animation and voice have arrived. Much more sophisticated examples exist in the world of computer games and the future holds undreamed of possibilities.

Interactive dial-in services are another wave of the future that has already gathered significant momentum. These services are computerized information and communication centers that you can connect to by using a personal computer, a modem, and your telephone. A modem is a box you attach to your computer that allows your computer to send and receive information through phone lines. A computer at the other end provides information and services. Dial-in information includes stock prices, news from wire services, weather and sports reports, and movie reviews, to name a few. Services include communication clubs that allow you to exchange messages with people of like interests and shopping services that let you buy anything from airline tickets to groceries.

According to Saul Carliner, in an article in *Technical Communication,* interactive video is another medium that will grow in future applications. This computer-controlled video allows the user to select the next sequence of images rather than to passively view whatever is on the video recording. Computer-controlled video is becoming more available as computer-readable media, like videodiscs and CD-ROM (Compact disc read-only-memory), become less expensive and more compatible with current computer technology. CD-ROM applications will allow you to play "What if?"—to rearrange your furniture or play "armchair travel."

"You can 'walk' down a street in a town," says Carliner, "stop, view the sites, take side trips, and look at available real estate—without ever leaving your computer."[11]

How these new and future media will impact the job of technical writing is still unknown. One thing, however, is quite clear. Technical communication is changing as rapidly as the technologies we document, and we'll have to stay alert and flexible to meet the challenges in our path.

 SUMMING IT UP

This chapter provided ways to organize technical information for both paper and online documentation. It described the document plan and scheduling as part of the planning process. It took a brief look at the future of technical communication media. The next chapter will describe the visual elements you'll control as part of the documentation process.

[1] R. John Brockmann, *Writing Better Computer User Documentation* (New York: John Wiley & Sons, 1986), pp. 51–52.

[2] Janice C. Redish, "Reading to Learn to Do," *The Technical Writing Teacher,* Fall 1988, p. 223.

[3] William K. Horton, *Designing and Writing Online Documentation: Help Files to Hypertext* (New York: John Wiley & Sons, 1990), p. 115.

[4] Janice C. Redish, et al., "Making Information Accessible to Readers," *Writing in Nonacademic Settings* (New York: Guilford Press, 1985), p. 142.

[5] Brockmann, pp. 206–213.

[6] Horton, chapters 1 and 2.

[7] Horton, pp. 24–26.

[8] Horton, pp. 16–24.

[9] Janice C. Redish et al., "Making Information Accessible to Readers," p. 144.

[10] Horton, pp. 106–107.

[11] Saul Carliner, "What's Ahead in Technical Communication?" *Technical Communication,* Journal of the Society for Technical Communication, Third Quarter 1989, p. 184.

11

■ ■

PLANNING FOR
VISUAL IMPACT

"I think there's a problem that writers tend to have," says writing consultant Daunna Minnich. "The reason people get into technical writing and stay in it is because they're logical, analytical people who are able to work with logical, analytical programmer types or technical types. And so all these people have real strong left brains, which is the logical, analytical part . . . But to be a good communicator—to be a good teacher— you need to appeal to more than just those logical abilities of people. So that's where it becomes important to try to find more than one way to get at an idea. This is where you get into the right brain, which is the part that perceives things spatially and is into music and art and all that kind of stuff."

The visual impact of your document is as important as the text. Yet often writers are not attuned to the elements of good page design and do not know how to visualize concepts very well. This is not surprising, as most writers have not been trained to think visually. Additionally, the visual elements of page design are very complex. Historically, they were handled by a professional designer.

Because page design and technical illustration have their own languages, some of the terms used in this chapter will be unfamiliar. If you see an italicized word you don't understand, refer to the "Glossary of Design Terms" at the end of this chapter.

127

▌ WRITERS AS DESKTOP PUBLISHERS

Now that desktop publishing software allows writers to format their text, many companies see it as a convenience to let writers do so. Typesetting is expensive and time-consuming and there's no reason to use it when the writer can supply the finished, camera-ready copy. This is true, but it skips a step. For typesetting, a manuscript first goes to a book designer who "specs the type"—describes the kinds of type and spacing that will be used for each kind of heading and text. The book designer has an eye for the subtle differences between *typefaces* and is sensitive to the impact a very small shift in spacing can have on readability.

With this expert out of the loop, document pages often look "MacTacky," as a colleague put it. When I asked what he meant, he said, "You know, they look like a ransom note."

What my acquaintance meant is that with a powerful desktop publishing tool, an inexperienced person can throw in too many different visual elements, just because the tool can provide them. The result looks amateur at best. Therefore, unless you have a design background, make use of graphic artists (or technical illustrators) whenever you can.

If you must design any aspect of a document, the absolute overriding principles to follow are these:

SIMPLICITY and CONSISTENCY

Now I'm going to seem to contradict myself. Within the bounds of good taste (which this chapter attempts to clarify), use as many visual elements as will enliven the page, making it interesting to the reader. But not just any visual elements. You will vary the look of your document with a controlled set of visual elements:

Typographical elements include:
- headings.
- text.
- labels on art.
- typographical emphasis to distinguish certain words.

White space includes:
- margins.
- indents.
- the space between text elements and between art and text.
- the space in illustrations.

Art is almost anything that's not text and for our purposes includes:
- charts.
- graphs.
- tables.
- pictures of screens.
- photographs.
- professionally drawn illustrations.

Each of these elements is controlled by rules, which I'll call the document's *format*. A document format is sometimes called a template, style sheet, or design specification. They all mean the same thing. Format dictates the kinds of type you'll use; the amount of white space between elements; whether art is boxed or floating, *line art* or *halftone,* and so on.

Now that desktop publishing is here to stay, we writers have got to hunker down and learn as much as we can about the visual elements of a book. And the visual part is a lot of fun.

AUDIENCE AND PURPOSE

In making visual decisions about your document, first consider its audience and purpose. The reading level of the audience and their patience with the material affect the size of type and the number and kind of illustrations you'll use. For example, an engineer is accustomed to pouring over long, grey blocks of text in relatively small type. While you can make his or her job easier by providing smaller paragraphs and clear, well-designed headings, you will probably provide a much more visually inviting page to the airplane mechanic, described in the chapter "Know Your Audience," who is covered with grease, working in the rain.

Says a graphic artist who illustrates repair guides for field engineers:

> More and more, I'm doing the art first, and some of the writers will take that piece of art and decide how to pare down the text. What they want to do is put as little text in as possible and only where it's appropriate. Especially in our manuals for field engineers who really like to get right to the point—to the purpose of the document. In a lot of cases, I've heard they even skip the text where they can and just go right to the illustrations. If they can get their info right away without having to read several lines, they'll do it.

▐ PAGE DESIGN

The impact of a page depends upon many things. Distinctions between levels of headings can be clarifying or confusing. The kind of type you use can give a clean or fussy look. The length of your paragraphs can invite or repel the reader.

Ultimately, if you know how to control the visual elements of the page, the page will look professional, creating a positive impression of your document, the product, and your company. Which is why large companies develop corporate design formats for their books. Companies like Xerox Corporation and Apple Computer go to great lengths to make their documents visually consistent. They're not trying to be dictatorial; they're trying to create a polished, corporate "look" that readers will recognize and trust.

If you work for a small- to medium-sized company, management might not have a clue about what goes into a document's design. It's up to you to package your writing as beautifully as your tools will allow or to contribute to your department's formatting decisions. The principles described in the following sections can help you.

Two excellent books on the general design principles of desktop publishing include Jonathan Price and Carlene Schnabel's *Desktop Publishing* and Daniel J. Makuta and William F. Lawrence's somewhat more technical *The Complete Desktop Publisher,* both listed in the bibliography. While newer books exist on the subject, these two provide good explanations of the basics.

WHITE SPACE

"White space is the space that is used but not printed," says Daunna Minnich.

> It's like, when you speak there's a lot of nonverbal communication. White space is the nonverbal part of communication in writing. It's the part that gives your eye and your mind some breathing space—some space to kind of invite you a little closer.

Leading, indents, and margins make up the bulk of white space on a page. The rest is provided by tables, charts, diagrams, and illustrations. Lists use more white space than regular text and thus invite the eye.

Studies have shown that if at least a third of the page is white space, text is easier to read. Even more white space is required for online documentation, where text appears to fill more space than on the page.

Leading

The leading between lines will vary with the typeface you use. Larger fonts appear to need proportionally less leading that smallers ones. Line length also affects leading. Research has shown that readability decreases as line length increases. You can compensate for this by increasing the leading between longer lines.

Experiment with different amounts of leading between lines of text. For example, put five short paragraphs on the same page, beginning with no leading and progressing to 4 *points* leading. Ask yourself if the lines appear too crammed together or too spacy. There'll be one paragraph for which the leading and type appear most balanced—somewhere between 1 and 4 points of leading for 9- to 12-point type. But this will vary with the style of type, so use your eye.

Headings should appear to belong to the material they introduce. Use about the same amount of leading before and after headings as between paragraphs, then adjust it so there's a little less space after the heading. That will separate it from the preceding material and lead your eye on to the next.

Margins and Indents

Margins can add white space to any format. Many technical manuals and commercial books now use an extra-wide left or outer margin to decrease line length and invite the eye. Major headings are placed on a *hanging indent,* out into the margin, where they are clearly visible for fast skimming.

Text can be either *right-justified* or *ragged right,* giving the page a very different look. Right-justified text lines up evenly at the right margin. Ragged right extends irregularly into the margin, giving the text a profile that helps readers recognize their place on the page. This makes ragged right more readable that right-justified text.

You will indent the entire left margin of some visual elements, so that additional white space frames them. Lists, notes, cautions, and art are good candidates for additional indenting. But be consistent. Says one technical illustrator:

> Writers need to align stuff. If they have a box, they need to align a margin with it. For example, when you're laying out a design for anything, you may

decide you'll have only three indents. Then everything will align with one of those three. It just gives a tighter, cleaner look. Then the eye knows where to look. Particularly in a technical manual, you want the reader to turn the page and not be surprised about the location of something. After a few pages, the eye wants to see something in the same spot on the page.

Art Spacing

Art placed within text needs space around it. The density of the art will make a difference in how much. If the art is very dark or detailed, it will require more space. Art itself can add a lot of white space to a page. If the kind of art you're using is light, place a box around it and use less space between it and text than you would for darker art. However, do not vary art spacing within a document. Once you decide on how much space to use between it and text, keep the same increments throughout. The density of the art should not vary either, but should have a consistent look, so the same spacing will work throughout.

Lists

Lists are a good tool for both visual and cognitive purposes. Short list items add sparkle to the page by providing extra white space and the staccato of bullets or numerals. Lists are usually formatted with a hanging indent, providing additional white space and breaking the left margin in a regular pattern. They invite reading.

Lists also organize material in bite-sized chunks the reader can rapidly grasp. They're the fast food of formatting. Therefore, when you mention more than two or three related items, try formatting them as a list.

TYPEFACES AND FONTS

Desktop publishing provides a vast range of contrasting typefaces from which to choose. "At first, explore these extremes," advise Price and Schnabel. "Later, you'll be able to find a quieter combination of typefaces and layout that gives your document a solid look of unity. As the poet Blake pointed out, the road to the Palace of Wisdom leads through excess."[1]

On my computer system, I can choose from 19 different typefaces, each of which comes in several sizes, bold face, and italics. Yet, to produce a manuscript, I use only one typeface in one font size. Even when I provide camera-ready copy to a client, I use only two or three typefaces: one for headings, one for text, and one for computer screens or examples. Again, simplicity and consistency are the guiding principles.

Select one typeface for your headings, another for your text, and stick to them. Distinguish between heading levels through alignment and typographical emphasis. (Typographical emphasis is described in the next section.)

Type comes in two basic styles: *serif* and *sans serif*. Studies have shown that serif type is slightly more readable, and I tend to favor it. But you can compensate for slower readability by using more leading with sans serif type, which some think has a more modern look.

Sans serif type provides a good contrast in headings, setting them apart from the body text. Sans serif is also more legible than serif type in very small font sizes for which serifs tend to blur. Use sans serif *callouts* in illustrations, where tight spacing limits the size and amount of type.

TYPOGRAPHICAL EMPHASIS

Within text, you'll use typographical emphasis to distinguish certain words from regular text. Below are some reasons you might use typographical emphasis:

- to emphasize a point (Do *not* touch the wire leading from the battery . . .).
- to distinguish the title of a book.
- to distinguish a term you are about to define.
- to indicate an example.
- to indicate a computer command the reader will type.
- to set apart error and system messages a computer might display.

Use typographical emphasis sparingly. Rarely use it to emphasize a point. If your point is important to the reader, set it apart as a note:

> **NOTE:**
> Log off the host before quitting the program. If you quit the program without logging off the host, the host maintains the session, and you will be charged for session time.

If your point could avoid damage to the product, set it apart as a caution, as shown on the following page.

> **C A U T I O N :**
>
> Make sure the mouse is attached before you turn on the computer. Attaching the mouse while the computer is running will seriously disrupt operation and may necessitate repairs.

If your point affects the reader's well being, set it apart as a warning:

> **W A R N I N G :**
>
> Do not touch the wire leading from the battery to the main board. You can receive a severe electrical shock.

Be consistent. For example, use italics for all new terms; do not switch to bold in a different chapter. Avoid using all capital letters. They grab attention and should be saved for the times you want to grab attention—like in a warning.

For examples of text to be typed into or displayed by a computer, indulge in a separate typeface or perhaps even in a separate color. Such examples are central to your reader's task and justify special treatment. If you have the budget, discreet use of a second color can be very appealing to readers. If you use a second color, be particularly shy of other visual elements. For example, use color instead of, rather than in addition to, a third typeface.

TABLES, CHARTS, AND GRAPHS

Tables, charts, and graphs are art that lie well within the writer's domain. Related lists can be formatted into tables. The progression of milestones in a schedule can be expressed in a flow chart. Percentages of a whole can be presented in a pie chart. Most relationships between two variables can and should be expressed in a graph.

As you write, repeatedly question whether your material can be better expressed in a table, chart, or graph. When a visual representation can convey meaning more economically than words, use it.

ILLUSTRATIONS

Illustrations can consist of only lines or a full range of shading. Figures 11-1 and 11-2 show two drawings of the same piece of equipment, one done as *line art* and the other in *halftones*.

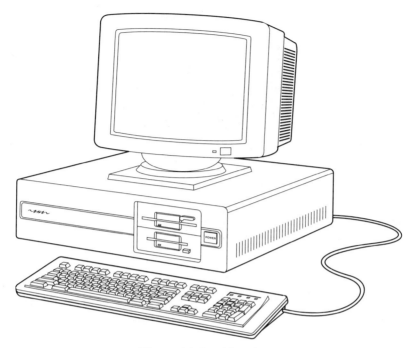

Figure 11-1.   Line Art

Illustrations can realistically describe a tangible product or abstractly represent relationships between biochemical agents or computer software products. Illustrations can be created by an artist using pen and ink or a computer graphics program. (The ones in Figures 11-1 and 11-2 were done by graphic artist Dany Galgani on a Macintosh IIx computer using Adobe Illustrator graphics software.)

You can take existing ink drawings or photographs and use a *scanner* to convert them into digitized art.

Illustrations can be used to simply entertain the reader, offering a pleasant visual change from text. However, most technical illustrations serve a more practical purpose: they communicate technical information. In this role, they supplement text or replace it completely.

If you think about how much time it would take you to verbally describe the location of every screw on a complex piece of machinery, you'll run to your nearest technical illustrator.

COMMUNICATING WITH ARTISTS

Artists can help you generate visual ideas. Approach them at the beginning of your project with a description of the document, the product,

Figure 11-2.   Half-Tone Art

and the kinds of illustrations you think might help. A staff artist keeps a library of published illustrations. Often you'll be able to select some drawings from past documents, which can be modified or used "as is" to illustrate your document. You'll provide the artist with a product specification and access to the actual product, so that he or she can modify old drawings or create new ones showing the most current version of the product.

Hardware products are easier to illustrate than less tangible products. For example, products like biochemicals or computer software require more imagination for you to conceptualize visually.

Says one artist:

I know so little about software that usually I get a sketch from the writer . . . They'll try to explain the principle of a sketch, and I'll come up with some ideas, as primitive as they may be, and then we start building from there. And the writer might say, no we can't do that, and then I'll come up with something else. It's like stepping stones—you go back and forth exchanging ideas. With the writer's greater understanding of the product and my understanding of how to present things visually, we usually work something out.

In addition to understanding the art you'll need, the artist might have to schedule your project along with many other writers' projects. Some of these will be higher priority than yours for marketing reasons. A "hot" product goes to the top of the pile. So while you're nervously waiting for the art to get into your document by your deadline, another writer may be breathing down the artists neck. Poor artist!

The best solution is to present the artist with a list of what you need early in the document schedule—as soon as you understand the product well enough to talk about it; then meet with the artist to jointly plan what needs to be done. At that point, the artist might ask to see certain equipment or blueprints you may not have provided or ask you to sketch something.

PLANNING ART

The following tips will help you plan the art for your document.

Tips on Planning Art

1. Read through your document outline and generate as many art ideas as you can.
2. Make up figure titles for the ideas and number them consecutively by chapter (for example, Figure 2-7 is the seventh figure in chapter 2).
3. Write your figure numbers and titles in two places: on a copy of your outline at the place where the figure will appear and on a separate figures list.
4. Look through related documents for illustrations you may be able to use, either as is or with modifications.
5. On your figures list, write the source document's title and the page or figure number, so the artist will know where to look for the existing illustration. If the staff artist is really on top of things, the illustration might have a serial number you can use.
6. Obtain blueprints or specifications for any piece of equipment the artist will need to draw.
7. Arrange for you and the artist to look at the piece of equipment as soon as a prototype is available.

8. If you want to illustrate an abstract concept, sketch first and go over your sketch with the engineer. Make sure your sketch is technically accurate before presenting it to the artist.

9. Give a copy of the figures list, specifications, your sketch, and any other information you've collected to the artist. Keep copies for yourself.

10. Make an appointment to discuss your document's visual and scheduling needs with the artist.

11. When you meet with the artist:

   - Ask any questions you may have about how best to illustrate concepts.
   - Discuss style issues: if the pictures will be boxed or left open, line-drawn or shaded. These style issues can affect your page design.
   - List any information or material you'll need to get for the artist.
   - Find out how the artist will supply the art. Will he or she give you computer art you'll insert in the document, or will the artist paste the art into your document? If the artist pastes the art in, how? By hand or computer? If the artist will insert art into your computer file from a graphics program, he or she will need both an online and *hard* (paper) copy of the finished document.
   - Discuss how much space to leave for each piece of art.
   - Discuss scheduling considerations: agree upon the deadline when you'll give the finished document to the artist for pasteup and the deadline when the document will be returned to you with the art pasted in.

12. Follow up on obtaining materials and information and on meeting your end of scheduling agreements.

Working with experienced graphic artists can be one of the most rewarding parts of your job, as your work will be greatly enhanced by their contributions.

 SUMMING IT UP

In this chapter, you learned about the visual aspects of page design and learned ways to work with a technical illustrator. In the next chapter, you'll explore the golden rules of expository writing and discover clear, concise ways to lead your reader through technical information.

GLOSSARY OF DESIGN TERMS

alignment
: Lining up visual objects, like a paragraph and a piece of art, so that they both share the same margin.

bold
: Type that is darkened for emphasis.

callouts
: Labels describing parts of an illustration.

flush
: Lined up against a margin. For example, a piece of art that is *flush left* butts against the left margin of the column.

font
: A *typeface* family member with specific properties. For example, a 12-point Times Roman bold font is a member of the Times Roman typeface family. *Font* is sometimes used synonymously with *typeface*.

format
: The page design, specifying type fonts, spacing, figures style, and so on.

halftone
: Shading produced by laying a screen of dots on areas of a drawing.

hanging indent
: Instead of being indented, the first line of a paragraph hangs out into the margin of the page. This list is an example.

justified
: Text lined up at a margin. Almost all text is left-justified, and can be either right-justified or ragged-right.

leading
: The space between lines. The word dates back to the time when typesetters used lead bars to separate lines of type.

line art
: An illustration drawn with lines, but without shading.

points
: The increment typesetters use to measure type and spacing. There are 72 points in an inch.

ragged right
: Text that is not right-justified contains differing line lengths, which create a ragged effect at the right margin of the page.

rule	A line, usually drawn across a page or column width.
running feet	A line of text, placed at the bottom of a page, usually consisting of a book, chapter, or section title and a page number.
running heads	A line of text, placed at the top of a page, usually consisting of a book, chapter, or section title and a page number.
sans-serif type	A plain type style, with no small extensions at the top or bottom of letters.
scanned art	Art produced by running an illustration or photograph through a *scanner*. A scanner *digitizes* the art, turning it into a file that can be read by a computer.
serif type	A style of type in which letters have small extensions at the top and bottom. This is an example.
typeface	A family of type designed to look alike; composed of several font sizes, which include punctuation marks, special characters (like parentheses), and numbers, as well as bold face and italics.

■ ■

[1] Jonathan Price and Carlene Schnabel, *Desktop Publishing* (New York: Ballantine Books, 1987), pp. 2–15.

WRITING IS THE HEART OF YOUR CRAFT

"They're fancy talkers about themselves, writers. If I had to give young writers advice, I would say don't listen to writers talking about writing or themselves."   Lillian Hellman, *The New York Times*, February 21, 1960

What is good writing? Good writing accomplishes a purpose. It does so economically and without apparent strain.

This chapter provides practical writing guidelines and explores some of the less tangible aspects of writing—developing style, finding your own rhythm within the writing process, and getting through writer's block.

If your background is more technical than literary, refer to the writing books listed in the bibliography for additional guidance. Writing after all is too great a subject to be taught in a single chapter.

 ## WRITING GUIDELINES

The rules of good expository writing prescribe simplicity, directness, and precision. You can apply them to any manual, journal article, training script, or for that matter, any nonfiction book.

This section describes writing guidelines as they apply to technical writing. You can use the checklist at the end of this section to review these guidelines while you write.

CHOOSE SHORT OVER LONG

Favor short, simple words over long, pretentious ones. Engineers and scientists tend to reverse this rule. A common example in technical writing is overuse of the noun *usage,* where the noun *use* can serve. Another is the verb *utilize* in place of the cleaner, simpler verb *use.*

If you can substitute one word for two, do. An example from business writing also corrupts technical writing: the use of *prior to* when *before* works just as well. Other evils include *in order to, as of now,* and their ilk. *In order to* can almost always become *to* and *as of now* is simply *now.*

Weed out unnecessary words by asking whether each word does a job. You'll soon learn to recognize suspicious words, like *very* and *particularly*—adverbs that rarely add to your meaning: The *particularly unusual event* is not much different from the *unusual event.*

In technical writing, short applies to sentences and paragraphs, as well. "Let us not fear lest we be too brief," advised turn-of-the-century tech writer Sir T. Clifford Allbutt. "If the matter be meagre padding will not amend it."[1] Confine your sentences to a single thought. This will keep them short and clear. Below is an example of a long sentence that has been rewritten as two shorter ones:

One long: As in the larger network, there is still a problem of interference by other traffic in the network, which may slow the flow in a particular path, or even stop it momentarily.

Two short: As in the larger network, interference from other network traffic still creates a problem. This interference can slow, or even momentarily stop, the flow through a particular path.

Short paragraphs invite reading. Presenting closely related material in short paragraphs is called "chunking." Technical writing research indicates that readers absorb material more readily in chunks, rather than through lengthy, ongoing discourse. As with single-thought sentences, construct paragraphs that center tightly on a single purpose.

CHOOSE ACTIVE OVER PASSIVE

An active-voice sentence contains a subject and an active verb. A passive-voice sentence contains an object and an inactive verb. The active sentence leaves no doubt about who is doing what to whom. The passive voice sentence both bores and confuses the reader. In the following sentence, who creates the text?:

Passive: When text is created with word-processing software, problems are caused by formatting commands such that machine translation is adversely affected.

Here's the sentence rewritten in active voice:

Active: When you use word-processing software to create text, the software embeds formatting commands that adversely affect machine translation.

You can recognize a passive sentence by "passive markers," says Roy Stewart in an article in *Technical Communication*.[2] These markers include a prepositional phrase beginning with *by* and sentences using the verb *to be (is, are, was,* and *were)* plus a past participle (verbs ending in *en, ed, d,* and *t*). Stewart advises:

> If writers train themselves to use these passive markers to test a passage after writing it, or—even better—while writing it, the markers would alert them to the passivity of such verb phrases as *is applied, was sent, is indicated,* and the like. Once alerted, the writers can make a conscious decision about the passive construction. If the sentence is passive and the decision is to leave it that way, I urge trying it in the active voice anyway. The new vigor and directness may overrule the reasons for leaving the sentence in the passive.[3]

Passive voice serves best in sentences that stress the verb or object, like the following:

Serious damage to the system can result if the wires are not connected in the correct sequence.

In this case, the reader's attention is rightfully directed to *serious damage* and *wires,* not to the system or the person connecting the wires.

Favor active voice; then use passive voice consciously, to emphasize the verb or object. When you use passive, though, make sure that the reader either does not need to know the subject of the action or is aware of the subject from a previous sentence.

CHOOSE SPECIFIC OVER GENERAL

General words leave readers confused. Specific words tell the reader exactly what's going on. Choose specific, precise words over general, vague ones. Here's an example:

General: If you don't type anything for an interval specified in the software, the system times out.

Specific: If you don't type anything for ten seconds, the terminal emulator program automatically logs off the host, and you must begin a new session.

The first sentence leaves the reader wondering: How long can I stop typing? What software? What does "time out" mean? What do I do if it happens? The second sentence provides concrete information the reader can understand and use immediately.

AVOID REDUNDANCY

Redundancy means unnecessary verbiage, usually through repetition. If you find yourself repeating instructions, definitions, descriptions, and illustrations, it means your document is not well-organized. This rule could have gone in the chapter "Organizing a Writing Project," but I include it here because redundancy shows up when you write.

If you find yourself repeating material, ask yourself if you can reorganize it so the reader doesn't need to read the same stuff twice. Are you repeating material to emphasize a point? If so, try to repeat the information in a different form, with a chart or table, instead of repeating text.

Just as you would avoid redundancy with explanations and illustrations, avoid redundancy with individual words. Beware of words that mean the same thing, like the ones below. Choose one word and trust it to do its job:

Redundant	*Improved*
assembled together	together
unrequired options	options
necessary requirements	requirements

ADDRESS THE READER

Historically, technical and scientific documents have been written in the third person, and some companies still insist their documents remain in this style. Third person refers to the reader as "the reader"; second person refers to the reader as "you." Third person sounds more formal and professional to some ears than the friendlier second person. But the drawbacks of third person far exceed its surface dignity.

In instructions, third person is unbearably tedious, as the following example illustrates. The imperative, in which "you" is implied, serves more effectively:

3rd person: The user then sets the switch by soldering wires 3 and 4.
Imperative: To set the switch, solder wires 3 and 4.

Addressing the reader directly, in second person or the imperative, enlivens your style. You'll write shorter, more active sentences and involve the reader in the action. Make sure, though, that you're clear about who your reader is. If your audience is programmers and you need to talk to them about the users of a computer program, then address the programmers as "you" and refer to the user as "the user"—third person.

DEFINE TERMS AND ACRONYMS

In the chapter "Know Your Subject," you listed technical terms and acronyms, and by the time you begin writing, you've researched their definitions.

You'll define technical terms and acronyms the first time you use them in your draft. If you use them often, the reader will remember what they mean.

If you define a term and then use it infrequently, you may need to redefine it with each use. This is particularly true if you're introducing the reader to a lot of new terminology.

One way to avoid cluttering your document with definitions is to include them all in a glossary; then tell readers to look there.

ELIMINATE JARGON

Jargon is the slang of a trade. Some jargon from technology has worked its way into everyday English. Examples include terms like *feedback* and *input,* which have come to mean response and advice respectively.

After working in a technical field for awhile, writers grow insensitive to its jargon. The more experienced they become, the more writers have to fight against jargon creeping into their work.

In most cases, jargon should not be defined, it should be eliminated. It is language that has developed out of laziness, convenience, or a desire to impress. A more meaningful expression can almost always be found. Play it safe—write English:

Jargon: Power up or reset the workstation.
English: Turn on the workstation or, if it's already on, press the reset button on its back panel.

USE PRESENT TENSE

In most technical documentation, use present tense verbs, even where you'd normally vary the tense. This convention seems awkward at first, but it's house style for most companies and readers have come to expect it. The following example illustrates this guideline:

Varied tense: If you select "yes" the program will display the next screen.
Present tense: If you select "yes" the program displays the next screen.

   I'm not fond of this convention, because it transgresses the rules of good English. The justification for it, as it was first explained to me, is that the reader is performing all tasks in the present and would be distracted by references to the past or future.
   You will develop an ear for present tense by reading technical manuals, and soon it will seem normal. Or I should say, you develop an ear for present tense by reading technical manuals, and soon it seems normal.

PROVIDE EXAMPLES

Examples almost always clarify technical material. Once you understand your subject well enough to write about it, you might conclude that your explanation is sufficient for the reader. However, because the subject is now familiar to you, your explanation might reach above readers' heads. An example will bring it down to earth again.

CREATE EFFECTIVE LISTS

Lists are critical in technical writing, both to describe step-by-step procedures and to group similar items in an easy-to-grasp, visually pleasing format. The effectiveness of a list depends on its

- consistency
- order
- completeness

In general, favor bullets over numbers. The preceding list is an example of a bulletted list. Use numbers with step-by-step instructions, items that will be referred to by number, and items introduced with a number. For example, the following list is introduced with a number:

Beginning programmers should be taught the following three rules:
1. Always list the objectives of the program.
2. Use a flowchart to diagram the program's operation.
3. Always include comments within your code.

Consistency

Items should be parallel in punctuation, capitalization, form, and content, so the reader can grasp their similarities easily. All items should begin with a capital letter or all with a lowercase letter. All should end with a period or without one. If one item is a complete sentence, try to make all items into complete sentences. Similarly, if one item is a prepositional phrase, all items should be prepositional phrases.

List items also should be parallel in content or meaning; they should be logically consistent. If you have difficulty making a list parallel, ask yourself if all items really belong there. How would you classify them? Your list items should be grouped together because they have some meaningful characteristic in common.

The following examples illustrate lists that are nonparallel and parallel in content:

Nonparallel content: Before configuring the software, you will need to:
1. Install the hardware.
2. Plug cable A into the connector on the back panel.
3. Configure the hardware.
4. Check that you have the correct software version.
5. Install the software.

Parallel content: Before configuring the software, you will need to:
1. Install the hardware.
2. Configure the hardware.
3. Install the software.

Order

Try to give your lists a logical order. If you cannot arrange items by size, importance, chronology, or some other logical order, arrange them alphabetically.

Chronological order is particularly important in step-by-step procedures. To ensure that the steps are in the correct order, use your document to perform the procedures. If you do not have access to the product, have someone like a field engineer do it for you.

Completeness

Completeness means everything is there. Like order, completeness is critical in step-by-step procedures, where a missing step leaves the reader hanging.

Completeness is also critical in lists of required equipment for a project or system. If a piece is missing, the project is held up. This can be expensive and maddening for customers, who will probably blame the documentation. Don't give them this excuse. Ask your technical resource person to check equipment lists for completeness, preferably by looking at the lists during a face-to-face interview.

AVOID ANTHROPOMORPHIZING

"The problem is not that it is difficult to make people accept a computer as human but that it is all too easy," says William Horton in his column "The Wired Word."[4]

Anthropomorphizing means referring to nonhuman entities, like computers, as if they were human. Programmers are particularly prone to this.

Beware of machines that *think, believe, assume,* and *conclude.* These are a few of the words programmers sometimes use to describe computer software. Usually you can replace these words with a simple action. For example,

Anthropomorphic:	After you stop typing for 10 seconds, the system assumes you're finished and blackens your screen.
"Computermorphic":	After you stop typing for 10 seconds, the screen turns black.

Some technologies have adopted anthropomorphic terms to describe, for example, certain software entities. *Users, clients, masters,* and *slaves* are all classes of data communication software, and their meanings are accepted in the field. Therefore, you cannot eliminate these words without depriving your reader. The best you can do is use them as adjectives: *user process, client process, master node,* and so on.

For online documentation, anthropomorphizing is particularly worrisome. William Horton describes a computer program named ELIZA, which was designed to function as a Rogerian psychologist. Its mimicry was so complete that, after a few minutes using it, even someone who knew the "psychologist" was only a computer program asked to be left alone in the room with it, so their privacy would remain undisturbed.[5]

Computer instructions that use the first person are cloyingly cutesy to experienced users. And ultimately, first person instructions break down, because the human model is inaccurate. The computer is simply not an *I*—it's an *it!*

"Don't pretend it isn't a computer," says Horton. "Help the user understand and anticipate how it operates." For example, in place of the computer message, "My memory is overloaded," Horton suggests the message, "Too many items to process."[6]

▪ ▪

**Checklist 12–1.
Writing and Revising**

The following checklist summarizes the guidelines presented in this section. Refer to it when you need help writing and revising your document.

- Choose short over long.
- Choose active over passive.
- Choose specific over general.
- Avoid redundancy.
- Address the reader.
- Define terms and acronyms.
- Eliminate jargon.
- Use present tense.
- Provide examples.
- Create effective lists.
- Avoid anthropomorphizing.

▪ ▪

WRITING STYLE

We usually use the expression *writing style* to mean the rhythm, tone, and various other intangibles that lend a piece of writing its character or effect. For example, words can sound formal or colloquial. Sentence lengths—short, long, or varying—provide rhythm. Each choice lends a particular quality to a piece of writing.

In technical writing, choices are limited, and style is dictated by the kind of writing you do, rather than by your preferences and personality.

In the chapter "Get Ready," you learned about four major kinds of technical writing: marketing communication and support; technical manuals and specifications; training materials; and technical journalism. In this section, you'll find examples demonstrating a writing style from each of these very broad categories.

Whatever kind of writing you do, develop an ear for the nuances of language, so that you can control your writing style. For example, an inexperienced writer might waver between a scholarly and chatty style, creating an amateur impression and distracting the reader from the document's purpose. Remember that anything that distracts from your communication goal does not belong in technical communication, or any good writing.

MARKETING COMMUNICATION AND SUPPORT

The goal of marketing communication and support is to convince potential customers to buy your product. Advertising provides the clearest examples of this communication goal. Here's part of an ad in *PC Magazine* for a computer board that speeds computer access to stored data:

- Want a 3- to 5-fold boost in system performance with no sacrifice of data integrity?
- Looking for a dramatic improvement in network productivity without replacing your 20, 25, or 33 MHz system or file server?
- Need faster data access? How about 0.3 ms? (That's over 30 times faster than the fastest hard drive on the market!)

Notice the choices that make up the ad's writing style: sentence fragments with no subjects simulate the verbal style of a salesman; technical terms offered without definitions assume reader sophistication; superlatives like *dramatic* and *faster* and a final exclamation point communicate emotional excitement found in no other kind of technical writing.

TECHNICAL MANUALS AND SPECIFICATIONS

Readers of technical manuals and specifications include both the technically sophisticated scientist and the uninitiated consumer. The readers are captive—that is, they have to read your document to understand a product or process. The following example is from an installation and operation manual for a hardware device that stores computerized engineering drawings, or *plots,* and sends them to a printer.

The RemotePlot handles both serial and parallel communication. To select whether output will be serial or parallel, use the serial/parallel switch, located on the RemotePlot front panel. When this switch is in the serial position, the RemotePlot outputs its data through the 9-pin serial port on the back of the unit. In serial mode, the RemotePlot can perform either XON/XOFF or hardware handshaking.

This paragraph's style is terse and direct. The writer doesn't try to draw the reader in. There's no need—the reader is actively installing the equipment while reading the material. The audience is engineers, so the writer doesn't define terms like *serial* and *handshaking*. Sentences are moderately long, but words are very short, making the material easy to read. Notice the complete lack of adjectives, emotionally colored words, or personal words. Every word serves a function. This is writing to inform.

TRAINING MATERIALS

The goal of training materials is to impart knowledge or skills that can later be applied to some task. Training materials are based on consciously formulated prerequisites and objectives, which are often stated in the beginning. The following example is from a training guide to be used with a textbook teaching System Network Architecture (SNA) fundamentals (SNA is a set of communication protocols developed by IBM):[7]

Audience   This course is designed primarily for programmers and other technical people who will be working with SNA data communication networks.

Prerequisites The recommended prerequisites for this course are:

- Introduction to System 370.
- Introduction to an IBM VS operating system.
- Knowledge of communications systems concepts, which include . . .

Objectives   At the completion of this course you should be able to:
- Define SNA terms.
- Identify the major components of an SNA network and describe their major functions.
- Describe data formats that are used in an SNA network . . .

After stating objectives, the guide provides exercises and answers that test whether or not the student has reached the objectives. Like technical manuals, this kind of writing is terse and direct, without adjectives or emotionally colored words. It is strictly formulaic: each section of the guide is parallel and organizational structures repeat. For example, each section contains an "Overview," "Objectives," "Required Materials," "Estimated Study Time," and so on. Terminology is stringently consistent.

TECHNICAL JOURNALISM

Technical journalism can be either scholarly or consumer-oriented. Scholarly journals present scientific or technical papers either ghostwritten by a technical writer or written by a scientist and revised by a technical writer to conform to a journal's style. Consumer publications, available through newsstands and direct mail, are closer relatives of other consumer magazines. They provide product reviews and how-to advice written in an informal style. Here's an example from *Light Plane Maintenance Magazine.*

> Replacing these cables is a perfect job for the aircraft owner who likes to work on his or her plane, since the job is more tedious than exacting, involving as it does removing and reinstalling seats, upholstery panels, and many little clamps, ties and knots. You'll have to have the job inspected and signed off by an A&P, but it's the sort of job it would be hard to screw up badly.[8]

The writer, Steven Lindblom, used a conversational style. His adjectives (*tedious, little*) and colloquialisms (*screw up*) lend a deceptively homey feel to the piece, which is very professionally written. He doesn't shy away from long sentences, yet his style flows seamlessly from long through short, creating a conversational rhythm. (You can tell I'm fond of this piece.)

This kind of technical writing allows the writer more freedom to develop a personal style than other, more restrictive forms.

THE WRITING PROCESS

The writing process differs greatly among writers. All technical writing, even the kind that tends to be formulaic and impersonal, is nonetheless a creative endeavor. As such, it relies on intuitive as well as mechanical solutions, and the writing process is affected by the writer's personality.

The order of the writing process described in this book should help your writing flow smoothly. You'll research your topic and audience first, then organize your document and plan its visual characteristics, and finally begin to write. It's a good order. However, no two writers are alike. You may need to begin writing first, as a way of exploring what you know and don't know about your subject. Some writers can't outline a complete project without writing parts of it first. (I confess I'm such a person.) Other writers outline the whole thing. "It's always the first and most important task," says writing consultant Linda Lininger. "And once I write an outline, I don't deviate from it."

So find your own way. Your first draft can be anything from a mass of scribbles to a finished manuscript. But no matter how carefully you write, edit your work to make sure it holds together and is mechanically perfect.

If the writing guidelines in this chapter are new to you, don't try to apply them all at once. Instead, write the best first draft you can, then use the guidelines to review the draft and rewrite it. Good writers sometimes revise a draft two or three (or more) times to perfect it.

ABOUT WRITER'S BLOCK

Writer's block besets writers for a number of reasons. One definition of writer's block is the failure of the writer's bottom to maintain contact with the chair.

"A writer will do anything to avoid the act of writing," says William Zinsser in *On Writing Well*. "I can testify from my newspaper days that the number of trips made to the water cooler per reporter-hour far exceeds the body's need for fluids."[9]

Causes of writer's block include:

- perfectionism
- inexperience
- fear of being wrong
- lack of information

The following sections describe these causes and suggests ways to break through to productive work.

PERFECTIONISM

Perfectionism can lead to writer's block. If you worry about getting every word perfect on the first pass, you might not get past word number one. A solution to this kind of block is to just write, even if it's garbage. You can delete the garbage later, but don't stop to judge the writing process.

INEXPERIENCE

Inexperienced writers suffer from a second kind of writer's block: sometimes they just don't know how to begin. They look at a blank page and think there's some mysterious process that gets the first word on the page. If you are such a person, consider—you use words every day to communicate, and spoken words are not that different from written words. If you don't know how to begin, try describing the product out loud, as if you were describing it to a friend over coffee. Then write it as you'd say it. Don't worry about the "well, sort of" and "I think it kind of." You can weed those out later. First just write.

FEAR OF BEING WRONG

Writer's block sometimes has an emotional component: you're worried about what others think of you or you're afraid of being wrong. In technical writing, you have to accept the facts that someone is always going to criticize your writing and that you will always get something wrong. Once you can live with those two realities, you'll probably get on with writing and never experience this kind of writer's block.

LACK OF INFORMATION

A fourth kind of block for technical writers comes from lack of information. If you try to write before you understand your subject well enough, you can experience great difficulty. Yet the lack of information is not always obvious. You may think you know enough to begin; then get stuck and not know why. Try asking yourself basic questions about your communication goal:

> **What do I want to accomplish in this chapter?**
> I want to describe how the product relates to other products the company sells.
> **O.K., which other products?**
> Well, there's a database that keeps track of engineering drawings, and . .
> **Good. How does your product relate to the database.**
> I think it has something to do with . . . I don't really know.

Using this questioning process, you'll find the holes in your knowledge. Write down your unanswered questions; then look through product specifications, use the product, or question the engineer until you understand the answers. The writing process will proceed more smoothly when you know what you're writing about.

▌ SUMMING IT UP

This chapter described principles of good technical writing, showed some ways writing styles vary, and discussed the writing process. The next chapter tells you how to edit and proofread your work, as well as how to communicate with editors.

▪ ▪

[1] Sir T. Clifford Allbutt, *Notes on the Composition of Scientific Papers* (London: MacMillan and Co., 1904), p. 14.

[2] Roy Stewart, "Writers Overuse the Passive Voice," *Technical Communication,* Journal of the Society for Technical Communication, First Quarter 1984, pp. 14–16.

[3] Stewart, p. 15.

[4] William Horton, "The Wired Word," *Technical Communication,* Journal of the Society for Technical Communication, Third Quarter 1989, p. 252.

[5] Horton, p. 252.

[6] Horton, p. 252.

[7] Luther Bonnett, *SNA Fundamentals: Personal Reference Guide* (Chicago: Science Research Associates, Inc.), p. v.

[8] Steven Lindblom, "Replacing Aluminum Battery Cables with Copper," *1991 Pilot's Guide to Maintenance and Operations* (reprints from *Light Plant Maintenance Magazine.* Greenwich, Conn.: Belvoir Publications, 1991), p. 4.

[9] William Zinsser, *On Writing Well* (New York: Harper Perennial, 1990), p. 22.

13

■ ■

EDITING YOUR WORK

E diting," said one writer, "saves you from making an ass of yourself. A good editor gets rid of your blunders before thousands of eyes can see them."

Editing imparts professional polish to your document. This chapter distinguishes between writing and editing and explains the importance of editing. You'll learn about different kinds of edits and find checklists to help you edit and proofread your work. You'll find guidance on dealing with a staff editor and on how to perform as one, if none is on staff.

 WRITING AND EDITING—WHAT'S THE DIFFERENCE?

Technical writing and technical editing share a flexible boundary. Writers are responsible for content; editors for polish. The many responsibilities in between are defined as writing or editing depending on the history of the publications department and its members' skills and backgrounds.

A technical document should receive at least three editing passes:

1. substantive editing
2. mechanical editing
3. proofreading

For our purposes, the definitions on the facing page will serve.

156

writing	selecting and organizing the content of a document.
substantive editing	revising to present material in the most usable, flowing manner.
mechanical editing	repairing flaws in grammar, usage, and punctuation and applying house style rules (later in this chapter, you'll learn about this kind of style); sometimes called *copy editing*.
proofreading	correcting typos and formatting errors; fixing disagreements between headings in the table of contents and in text, in the index and in text, and the like.

Substantive editing forms the gray area between writing and editing and usually lies within the technical writer's domain. This chapter concerns itself primarily with mechanical editing and proofreading.

 ## WHY EDIT?

Editing increases readability. One way it does this is by treating similar elements consistently. For example, if a marketing brochure refers to a product as Power Tool, an editor can ensure that the reference manual and online information for the product also refer to it by the same name, spelled and capitalized the same way.

This consistency prevents customer confusion and also gives the product a "corporate look." When terms are used consistently, customers easily recognize them. And product names and terms, like trademarks, gain loyalty through recognition.

Consistency consistency consistency. This chapter emphasizes it because consistency is almost synonymous with editing. You might get tired of seeing the word, but it's repeated for good reason.

Consistent terminology is only one reason to honor the editing process, but it's a good example, affecting as it does the total impression the product creates. Other mechanical repairs to a document also influence product acceptance. Would you trust a product if you found numerous typos in the manual?

You can see that polished documentation reflects product quality. Now let's explore the three editing passes.

SUBSTANTIVE EDITING

When you perform a substantive edit, you might arrange material more logically to eliminate redundancy and make the document easier to use.

You'll substitute specific words for vague ones and redefine terms to inject clarity. A good substantive edit requires some technical knowledge, and rightly rests in the writer's hands, although some writers do not have the skill required. If a document is written by a subject matter expert, like an engineer, the responsibility for structuring the material might fall to an editor.

The principles of substantive editing are the same as those for good writing. Therefore, to perform this kind of edit, you can follow the writing guideline described in the chapter "Writing Is the Heart of Your Craft."

 MECHANICAL EDITING

Mechanical editing lies squarely on the technical editor's turf, provided you have such a person on staff. If you do not, you'll need to do it yourself.

When you perform a mechanical edit, you'll apply the rules of grammar, usage, capitalization, and punctuation. In addition, you'll check that technical terms are defined and are consistent in spelling and in the way they're used. You'll make sure like items are parallel. And you'll enforce *house style,* which dictates the spelling, punctuation, and typographic emphasis of product-related terms. The elements of house style and use of style guides are described in the section "House Style," later in this chapter.

Whether you edit your own document or work with an editor, learn to recognize and use the editing marks shown in Figure 13-1.

A complete treatment of grammar, usage, punctuation, and consistency is beyond the scope of this book. The following sections list a few rules writers frequently abuse.

For more details about punctuation and word treatment, as well as editing marks, see *The Chicago Manual of Style,* published by the University of Chicago Press and frequently revised. If you plan to edit much, buy a copy.

AGREEMENT BETWEEN SUBJECT AND VERB

When a singular subject is separated from the verb by a plural noun or a series of nouns, the verb should be singular. Writers sometimes use a plural verb, creating subject-verb disagreement. An example follows.

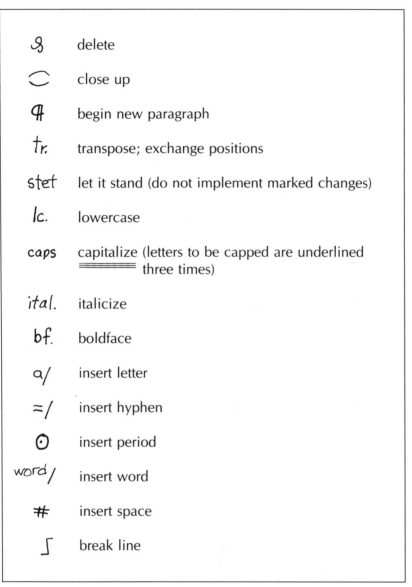

Figure 13-1. Editing Marks

Disagreement: The graph showing dates, sources, and responses appear on the facing page.
Agreement: The graph showing dates, sources, and responses appears on the facing page.

The ear tends to hear a plural verb after plural nouns. However, the verb *appear* refers to *graph,* and not the series preceding the verb.

COMMAS IN A SERIES

When you write a series of nouns or phrases within a sentence, you have a choice about whether or not to place a comma before the *and.* That last comma is called a *serial comma.* For example:

No serial comma: Before opening the case, make sure you have on hand a 1/4-inch screwdriver, a 3/4-inch wrench and some wire clippers.

Serial comma: Before opening the case, make sure you have on hand a 1/4-inch screwdriver, a 3/4-inch wrench, and some wire clippers.

Technical manuals almost always use serial commas. The important thing is to be aware of your company's style preference and to be consistent: either always include a serial comma or never include one.

COMMAS AND PERIODS WITH QUOTATION MARKS

A comma or period should appear before the final mark in a quotation, rather than after it. For example:

When you reach the heading "Software Agreement," begin reading carefully to make sure you understand the terms.

However, the computer industry has corrupted this rule. This corruption began when writers placed quotation marks around computer commands. If a comma or period appeared within the quotes, the user was likely to type it as part of the command. If the punctuation mark truly wasn't part of the command, it fowled things up. So rightly, tech writers moved all extra punctuation outside quotes, leaving within them only those characters the computer could recognize. Since many writers weren't quite sure of the rule anyway, commas and periods started creeping out of quotation marks in all instances.

A comma or period belongs inside quotes unless it will confuse the user. One way around the computer-command problem is to use typographical emphasis, like italics or bold type, to set commands apart, rather than placing them inside quotations.

If your company uses quotes when describing commands and online messages within a manual, do not use commas or any other characters within the quotes except exactly what the user must type or what appears on the screen. In other text, however, place these punctuation marks within the quotes where they belong. For example:

Computer command: After you type "HELP", a list of topics appears on the screen.

Other text: The resin-curing process is "air-inhibited."

CONSISTENT PERSON

Refer to the user consistently as either "the user" or "you." Do not switch between second and third person in your document. For example:

Inconsistent: The purpose of this manual is to help you operate the XYZ machine. Users can learn all XYZ features easily by following the step-by-step procedures in chapters 2 and 3.

Consistent (third person):   The purpose of this manual is to help users operate the XYZ machine. Users can learn all XYZ features easily by following the step-by-step procedures in chapters 2 and 3.

Consistent (second person):   The purpose of this manual is to help you operate the XYZ machine. You can learn all XYZ features easily by following the step-by-step procedures in chapters 2 and 3.

CONSISTENT CAPITALIZATION

First, some editing terms—When editors tell you to *capitalize* a word, they usually mean only the first letter. *Initial cap* also means the first letter in a word is a capital. All others are lowercase, like so—Aardvark.

All caps means all letters in the word are capitalized, like so—AARD-VARK.

In general, favor lowercase letters and use capitals only to distinguish proper nouns, like the name of a product. Engineers tend not to know this rule and can randomly capitalize words in specifications and online. This practice is particularly troublesome when you're trying to create consistency between your document and computer screens designed by engineers.

House style sometimes dictates that commands and file names be capitalized, although other forms of typographical emphasis work just as well. Boldface, for example, can be used to indicate a command: **exit.**

Educate your engineers and fellow writers to use capitals only for proper nouns and for their usual roles in titles, captions, and headings, and at the beginning of sentences and some list items.

If you do capitalize an element like command names, make sure to do so consistently throughout the product documentation to avoid confusing your audience.

Capitalization of Titles, Captions, and Headings

Capitalize titles, captions, and headings consistently. Titles, captions, and headings generally follow these rules: Always capitalize the first word. Use lowercase letters to begin prepositions (*to, with,* etc.) articles (*the, a, an,* etc.), and conjunctions (*and, or,* etc.). Capitalize all other words. The heading for this section is an example.

Follow house style, which may differ from this rule. Then edit all titles, captions, and headings to ensure that they conform to the same capitalization style.

Capitalization and Punctuation of Lists

Capitalize and punctuate list items consistently. For bulleted lists, capitalize or lowercase the first word in each list item, depending on house style. You'll almost always capitalize the first word and place punctuation at the end of list items that form complete sentences. Here's an example from the "Resumes" checklist earlier in this book:

- Have you listed your technical skills?
- Is the layout attractive?

Often you'll lowercase list items that are nouns or that complete a sentence begun by the introductory phrase. End punctuation is optional. However, house style might specify end punctuation after the items if they complete a sentence.

The following example shows lowercase nouns listed with no end punctuation:

- one external disk drive
- one internal hard disk
- two coaxial cables

Again, consistency is the goal—make sure similar lists are treated similarly.

PARALLEL WORDING

Phrase list items and headings in parallel form. For example, if one heading contains a gerund (the verb form ending in -ing), use a gerund for all similar headings of the same level in the same section.

The following examples show nonparallel and parallel headings:

Nonparallel: Planting for Maximum Air Circulation
                The Use of Insecticides
Parallel: Planting for Maximum Air Circulation
                Using Insecticides

Similarly, edit list items to be parallel, as described in the chapter "Writing Is the Heart of Your Craft."

PROOFREADING

Proofread every draft you send for review, as well as the final camera-ready copy. After you type in corrections, proofread your changes to make sure you haven't introduced new errors.

When you proofread, you'll correct spelling and typos, check headings and page numbers against the table of contents, check index entries against pages to which they refer, and so on. (The principles of indexing are described in the chapter "The Production Process," because the index is the last thing you'll do—and the last thing you'll proofread.)

If you do not have an editing or production staff, you'll proofread formatting elements to ensure that they've been implemented consistently. You'll check things like type fonts and sizes, as well as spacing.

Some of these proofreading tasks belong to the final phase of document production and are described in the chapter "The Production Process."

Use the proofreading checklist later in this chapter to remind yourself of elements that need "proofing."

SPELLING AND TYPOS

Most technical writers use a word processing program that includes a spelling-checking function or they use a separate *spell checker*. Spell checkers are computer programs that refer to a standard dictionary, stored in the computer's memory, to check the spelling of every word in your document. Most of them do this very quickly—faster than you can grab your dictionary. Nonetheless, they don't catch all errors. Here are the common ones spell checkers usually miss:

- double words, like "the the" (some spell checkers catch these)
- incorrect words spelled correctly, like "than" where you meant "then"
- incorrect numbers of spaces between words and sentences
- product-related terms and acronyms

Therefore, despite the aid of this high-tech tool, you need to proofread your document for spelling errors and typos.

CROSS-REFERENCES

Proofread all cross-references. If the text states "see page 9 for a description of binary notation," turn to page 9 and make sure the description is there. To reduce the risk of incorrect cross-references, refer to chapter or section names, rather than specific page numbers. References to page numbers are easily rendered inaccurate because as you edit a document, you delete and add material, thereby changing where one page ends and the next begins. Material that fell on one page soon falls on another. Therefore, use names rather than page numbers, unless house style dictates otherwise. And proofread cross-references, the table of contents, and the index at the end, when pagination is finalized.

FORMATTING

Proofread headings, captions, running heads, and running feet to ensure their consistency. Make sure margins, indents, and leading look consistent around parallel elements.

Similarly, scan text for font errors. Word processing can introduce stray fonts and paragraph styles, and in your hurry to meet the deadline, you don't notice them. When you proofread, stray fonts will appear somewhat darker or lighter than the text around them, but you have to look closely for them.

As you proofread, look for paragraphs that are incorrectly indented and spaced; they may be formatted as list items, headings, or in some other style. Similarly, proofread other text categories, such as list items, to ensure they are formatted correctly.

Use the Proofreading checklist to remember those elements that require proofreading.

■ ■

Checklist 13–1.
Proofreading

1. Look for spelling errors and typos your spell checker missed.
2. Check for correct numerical order of numbered lists and procedures.
3. Check for correct alphabetical order of alphabetized lists.
4. Verify cross-references for content.
5. Make sure spacing and type style are consistent for all headings and captions of the same kind, including running heads and running feet.
6. Check headings (including chapter heads) against the table of contents to ensure they agree in level, wording, and capitalization.
7. Check captions against the figures and tables lists for agreement.
8. Check text for stray fonts.
9. Check art placement (described in the chapter "The Production Process").
10. Look for widows and orphans (described in the chapter "The Production Process").
11. Proofread page numbers in cross references, the table of contents, figures list, and tables list.
12. Check index entries for correct alphabetical order.
13. If you've used a computer program to create the index, spot check index page numbers to ensure that they refer to correct pages. For a manually created index, proofread the whole thing.

■ ■

HOUSE STYLE

House style is a set of rules that govern the treatment of product-related terms and of elements, like lists, for which no universal rules exist.

Consistency (that word again) is the heart beat of house style.

You've already learned about some of the decisions dictated by house style—decisions like how to emphasize command names and whether or not to capitalize list items. Formatting elements, like fonts and margins, have also been mentioned.

House style also includes rules about the spelling, hyphenation, and capitalization of product-specific terms. For example, a computer manual needs to present commands, keystrokes, and online messages in consistent style.

Most technical publication departments use a style guide to ward off the confusion engendered by multiple representations of like elements. The style guide specifies one correct way to present terms that might otherwise appear haphazardly. The following is an example from a style guide for computer manuals.

Capitalization and typographic emphasis for technical terms:

- Show commands lowercase and boldface, unless the command must by typed in capital letters.
- Show product labels capitalized as they appear on equipment.
- Italicize file names.
- Use initial capital for program names.

If the responsibility for style decisions falls in your lap, as it may in a small company, you need to produce at least a style sheet, which you can refer to and pass to other writers working on the same product documentation. You may even need to create a style from scratch. The sheet or guide should specify how to handle the elements listed in the following checklist. The list is not complete. You will constantly update your style reference to include new elements as you find them.

▪ ▪

Checklist 13 – 2.
Style Elements

When you create a style guide, you'll decide the following issues:

- Capitalization, font, typographical emphasis, and spacing of chapter titles, headings, and captions.
- Number of heading levels to use.
- Capitalization, typographical emphasis, and spelling of product-related terms, like commands.
- Typographical emphasis and placement of notes, warnings, and cautions.
- Capitalization, spelling, and punctuation of measurements, like weights and rates.
- Whether to abbreviate or spell out measurements.
- Capitalization and punctuation of acronyms and abbreviations, as well as their definitions.
- Punctuation and division of terms that can be represented as two words or one, for example, *log-in, log in, login.*
- Placement and punctuation of table and figure captions.

- Capitalization, punctuation, and indentation of list elements and index entries.
- Use of serial comma.

▪ ▪

You will also need to choose a dictionary and style guide to use for more general spelling and style decisions—those that plague everyone rather than those that apply only to your company. *The Chicago Manual of Style* is the style guide preferred by most technical publications departments and is an excellent reference to use in preparing your company-specific style guide.

WHO EDITS?

The roles of technical writers and technical editors are not sharply delineated, and in the business world, these two job titles are sometimes used interchangeably. For example, when I was employed as a technical editor, I sometimes rewrote technical material, whereas in one technical writing position, my work was confined primarily to mechanical editing. As a technical writer, I confronted my first blank page, to be filled with original writing. However, I've met people with the title of editor who also do original writing.

Be aware of your preferences. Some technical editors want only to edit. That is, they do not want to have to know about a technical product or to be responsible for the technical content of a document. Similarly, some writers (I, for one) prefer the intellectual challenge that comes with learning technical information and structuring it into some form of communication.

If you have a strong preference for either editing or writing, do not mistake a job title for a job definition. Find out what a title means at the company you're approaching for a job.

NEGOTIATING WITH AN EDITOR

If you have both writers and editors within your publications department, the editor will edit your draft. If polish is a priority at your company, the editor may go through your draft as many as three times, at different stages. First he or she might recommend organizational changes; a second, thorough mechanical edit might come next; and during a third edit, the editor will check that you've implemented mechanical corrections and will proofread formatting, table of contents, and so on.

"Take confidence from the fact that almost any editor worthy of the name can improve the work of almost any writer, however talented," says Jefferson Bates in *Writing with Precision*. "The reason? A 'cold eye.' This is especially true when the original writer has drawn information from source materials of dubious quality—badly organized, poorly written, or not factual. Some of these faults almost inevitably rub off; a second writer or editor taking a fresh look has the immediate advantage of being one step further removed from the original garbage."[1]

Some writers see editors as persecutors, to be argued with at every turn. Others lean too heavily on them, submitting rougher drafts than an editor can be expected to rescue. Striking a balance between these two positions is best. Editors are your colleagues. Particularly in technical writing, their job is not that different from yours. They will make mistakes, but they will also help you enormously to produce a professional publication.

You are responsible for letting editors know when they have inadvertently compromised technical accuracy. You should have the final say about content matters. On the other hand, don't leave all the cleaning up for the editor. The more you edit your own work, the fewer red marks will appear on it during the editing process. And your attention to the mechanical details of a manuscript shows editors you value what they do. Editors appreciate your efforts, and your work relationships with them will go more smoothly.

PEER EDITING

If your department does not define a separate editorial function, you will edit your own work. To provide consistency among a group of writers, members must agree to follow a shared style. This requires a team effort.

When I began as senior writer at one company where no staff editor existed, I found the company's manuals contained a hodgepodge of styles. Product-specific terms varied haphazardly in their spelling, capitalization, and punctuation. I found as many as seven different representations of a keystroke sequence in a single manual.

I offered to help clear up the confusion and organized a weekly usage panel, which writers voluntarily attended, to decide style issues. Together we divided up issues, researched style guides and good manuals from other companies, and brought recommendations back to the group.

The writers all felt involved, yet there were drawbacks. When it came time to implement the recommendations, writers who didn't attend the panel or hadn't researched a particular issue, continued in their old ways.

For whatever reasons—deadline pressures, habit, lack of managerial fol-low-through—inconsistencies continued to appear.

This experience taught me to respect the role of a staff editor. Some-one who's job it is to keep track of the nits will do so with greater diligence than a writing team can possibly afford to do. Nonetheless, a team effort is better than no effort at all.

SELF-EDITING

Without a staff editor, and sometimes even with one, you are ultimately responsible for the consistency and polish of your work. If you have little time, make every effort to edit your draft at least once, even if it means taking it home for the weekend. Prefer at least three passes: substantively edit, mechanically edit, and proofread the draft. Ensure that, at least within itself, the document is consistent and that it is error free.

 SUMMING IT UP

This chapter described the editing process and the value of consistency. It guided you in editing your own work, in deciding style issues, and in working with a staff editor. The next chapter tells you how to prepare for the review process, how to submit your work to reviewers, and how to use their comments to improve your work.

▮ ▮

[1] Jefferson D. Bates, *Writing with Precision* (Washington, D.C.: Acropolis Books, Ltd. 1978), p. 11.

14

■ ■

THE REVIEW PROCESS

To be accurate and useful, your work must be reviewed several times by experts from different areas of your company. During this process, you'll learn to take criticism, to incorporate technical changes consistently, and to reconcile differences among reviewers.

 THE MYTH OF EGOLESS WRITING

Your technical document will pass under many often unsympathetic eyes during the review process. I've had more than one acquaintance ask, "How can you stand it?" when they hear about this aspect of technical writing. How do you survive the review process? Experienced writers recommend that you get your ego out of the way. However, pure egoless writing is a myth.

If you care about your writing and strive to make the best choice you can in its form and content, you feel bad when a reviewer says your work is garbage. Very few writers get their egos out of the way completely. Often those that do are simply burned out, and their poor-quality work reflects their lack of ego involvement.

TAKING CRITICISM

Learning to take criticism less personally takes time and is never complete, because it is your ego—your pride and confidence in your work—that guides you in dealing with reviewer's comments.

One way to take criticism less personally is to realize that your document is a company product. Others have vested interest in it, and as a team player, you're responsible for soliciting their contributions. In doing so, you need to cultivate a corporate, rather than personal, perspective on your work.

"When I started in the field, having never been seriously challenged about my writing, it was very difficult," says biomedical writer Dan Liberthson. "I got my back up for awhile until I realized that [the review process] was valuable."

You'll find after awhile that your problem won't be how to take negative review comments but how to convince reviewers to make more of them. Because you depend on product experts to make your document as useful as possible, you'll look for thorough reviewers—as many as you can get—and you'll grow impatient with those who send your draft back with few comments.

How do you discriminate between useful and hurtful review comments? If you incorporate every criticism, your writing loses its coherence. If you reject every criticism, you lose opportunities to improve your writing and you become known as a difficult person. Worst of all, you risk the technical accuracy of your document. Taking criticism demands that you balance subjectivity and objectivity; respect for your own point of view and that of others. It's an art.

The following checklist will help you sort through changes reviewers request.

▪ ▪

Checklist 14 – 1.
The Review Process

Apply the following criteria to reviewers' comments to determine whether they help or hinder your communication goal.
  Make a requested change if it:

- improves the technical accuracy of the document.
- improves the organization of the document, making information more accessible.
- clarifies an explanation or procedure, making it easier to understand.

Do not make the change if it:

- introduces material irrelevant to the stated audience.
- compromises the document's mechanical correctness, coherence, or readability.
- conflicts with other reviewers' comments or with technical accuracy as you understand it. (Such changes need to be discussed with reviewers, as described in the section "Incorporating Review Comments" later in this chapter.)

WHO ARE YOUR REVIEWERS?

Reviewers are usually selected from each of the following departments:

- marketing
- engineering
- quality assurance (QA) or testing
- customer support
- publications
- editing

Your company might have different departments than the ones listed here, and representatives from those departments, or from other branches of the company, will review your work. The following sections describe typical reviewers, so you'll know the kinds of comments to expect and how to use them.

MARKETING

If you're writing directly for a marketing department, marketing reviewers will comment on your document as an effective sales tool. If you're writing for a different department, reviewers from marketing will help you coordinate your terminology with theirs. They will probably try to delete any negative references to the product. Be careful about making these deletions, as they can deprive the user of necessary information.

Make sure that in response to a marketing review you do not delete information warning users about hazards to themselves and their equipment. Also state positively, rather than delete, descriptions of bugs engineers could not fix before the product was released. For example, if a computer command that appears on the screen does not work yet, let the user know it "will be implemented in the next release."

ENGINEERING

If you are writing technical manuals or specifications, you should give review copies to several engineers. If you are writing other kinds of material, an engineering reviewer is still essential. Engineers are closest to the product and most able to ensure the technical accuracy of your work.

QA AND TESTING

QA and testing professionals use the product to ensure that it runs smoothly. Try to have them review the manual by testing it alongside the product. They can determine if your written procedures really work, and their comments are invaluable.

CUSTOMER SUPPORT

Customer support engineers (including field support and sales support) help set up the product and ensure that it runs smoothly at the customer site. Your company might also employ customer support engineers to do telephone troubleshooting and on-site repairs when customers have difficulty operating the product. Reviewers from customer support not only help with technical accuracy, but also help you define the audience for your document. They know the customer. They may inject different terminology, perhaps even jargon, that the customer regularly uses to refer to the product. You might need to include these terms (in addition to company standard definitions) to reach your audience.

PUBLICATIONS

Sometimes a review copy will go to a writer in your publications group, or a lead writer, for a peer review. Usually your supervisor will receive a copy. A writing reviewer can help with organization, writing style, and clarity.

EDITING

The editor will review your document for mechanical correctness, house style, and consistency, as described in the chapter "Editing Your Work."

OTHER REVIEWERS

A reviewer from the training department can also provide helpful recommendations about a range of concerns, including terminology, technical content, and document organization.

In addition, a review copy should go to the artist, who can use it to anticipate art placement and recommend changes affecting art. And inevitably you'll send out some "political" review copies to people who never have time to read your work but nonetheless reserve the right to review it. These are usually engineering and marketing managers, and you'll need to deal with them more delicately than reviewers who are actually expected to return comments to you by a deadline.

 ## ASKING FOR TROUBLE—SUBMITTING AND RETRIEVING THOSE REVIEWS

You will send your work out for review after you complete the first draft and again after you have incorporated reviewers' requested changes. Sometimes a third review is necessary, if reviewers have requested extensive changes or if the product has undergone changes.

When you submit your work for review, you'll include a cover memo with your name and phone number, the name of the document, the names of reviewers, and the date reviews are due. Distinguish between critical and "FYI" (for your information) reviewers. All your main reviewers are critical. The most critical is your primary technical resource person, whose comments help ensure your document's accuracy.

Your artist, your manager, and the political reviewers mentioned earlier are less critical. Unless you write training materials, training reviewers should also be FYI, rather than critical, because waiting for their comments might hold you up—trainers are often out of town.

Figure 14-1 shows an example of a review cover memo.

Only one task is more difficult than submitting your work for review, and that is getting it back from late reviewers. It's hard enough to ask for trouble in the first place. Having to beg for it is a disgrace!

When writer Lisa Peters (fictional name, but real incident) asked one late reviewer when she would receive his comments, he answered, "When pigs grow wings." That afternoon, Lisa bought colored construction paper and glue, and created a mobile for this engineer. When he returned to his office the following morning, he found flying above his desk a winged pig.

Lisa's creativity paid off, but it also cramped her deadline-ridden schedule. Late reviews can cost hours of a writer's time and can delay the manual's publication date. At meetings on the subject, writers brainstorm such ideas as attaching lottery tickets or five-dollar bills to reviews as

---

Figure 14-1.   Review Cover Memo

March 23, 1992
TO: Reviewers
FROM: Janet Van Wicklen, ext. 897
RE: Second review of *The Wild Widget User's Guide*

Please review the attached document for technical accuracy. Comments not related to technical content are welcome but might not be incorporated. I need your comments by **Friday, March 31st.** Critical reviews received later than this date will jeopardize the document's schedule. If you will not be able to review the document by this date, please call me immediately.

Please write corrections directly on the draft. Make comments specific: If you mark something as wrong, also write in the corrected information.

Thanks for your help making this an accurate, user-friendly document.

REVIEWERS:

Critical Reviewers	*FYI*
Susan Black, Engineering	Harold Schwartz, Engineering
Grant Smith, Engineering	Jane Townsend, Marketing
Lisa Downing, Marketing	Betsy Long, Technical
George Freeman, QA	Publications
Pete McMann, Customer	John Ferrero, Technical
Support	Publications
Pam Brown, Technical	Jim Hartford, Training
Publications	

---

bribes, or including quasi-humorous threats like "REVIEW THIS OR ELSE." The best solution is to convince reviewers' managers that document reviews are an essential part of the production process and that late reviews can hold up the product.

In your review cover memo, clearly state the date beyond which the manual's schedule will be jeopardized if reviews are not returned. If your management agrees, inform reviewers that the publication date will slip a day for every day critical reviews are late. This works well because no one likes to be seen as the bottleneck in the production process.

 ## INCORPORATING REVIEW COMMENTS

You will receive many incomplete, illegible, or confusing review comments, despite your request for specific information. Reviewers either don't read your cover memo or do not know how to clearly write the correction. One common, maddening review comment is "See me"—maddening because you are under deadline pressure, have a half dozen or so reviewers' comments to coordinate, and cannot possibly sit down with all of them to clarify their concerns.

Nonetheless, try to follow up on incomplete review comments. You'll quickly learn who your helpful reviewers are—those who know the product best and who know how to describe its intricacies. You'll take more time clarifying their comments than those of less-experienced reviewers.

CONSISTENCY

Apply requested changes consistently. Changes to names, parameters, commands, filenames, procedures, and product descriptions need to be made throughout the product documentation. If a reviewer marks a change in one part of your document, make the change wherever else it applies. For example, if a procedure appears in both the user's guide and the quick-reference card, and a reviewer marks a change in the user's guide, take responsibility for changing it in the quick-reference card as well. To do this may mean you have to team up with other writers in your department.

CONFLICT RESOLUTION

Inevitably, two or more reviewers will disagree about how the product works. Try to convince reviewers over the phone to resolve their differences and get back to you. Often you can ask your technical resource person to clear up the confusion. If resolution isn't possible, host a review meeting, inviting the disagreeing parties to bring their comments for discussion. Some publications departments routinely hold review meetings to enable document reviewers to agree on changes face-to-face.

SUMMING IT UP

This chapter described how to approach the sometimes harrowing review process and use it to improve your document. The next chapter describes your document's graduation from a rough draft to camera-ready copy.

15

■ ■

THE PRODUCTION
PROCESS

Think of the production process as gift wrapping. You've done everything you can to create a finished document, but you still need to package it to present to your reader. This stage involves polishing your format, proofreading the final draft, and creating that vital part of your document—the index. Finally you will print out a camera-ready copy to send to the printer.

When you've completed these steps, you'll admire your work, update your resume to reflect your new accomplishment, and celebrate a job well done.

Large companies sometimes have a production staff that handles this phase. In most companies, however, you will be responsible for the camera-ready copy.

▍ POLISHING YOUR FORMAT

The manuscript you send to the printer should be perfectly formatted, with final placement of art, nicely balanced pages, and no *widows* or *orphans*.

PLACING ART

Changes you made during the review process may have affected the placement of tables and figures. Perhaps some of them now precede the text explaining them. This can confuse the reader, who will wonder why the art is there. Now it's time to proofread the text around tables and figures to make sure that the text refers to them before they appear—not after.

BALANCING PAGES

Next, proofread running heads, running feet, and page numbers to make sure that they align horizontally across facing pages and vertically with margins. Also, check that even and odd pages mirror each other. Page numbers should appear on the outer edge of the page, so the reader can flip through your document to find his or her place. Therefore, the page number should appear on the left side of an evenly numbered page and on the right side of an oddly numbered page, although some formats might specify that page numbers be centered. Similarly, running heads or running feet should also mirror each other.

If your format includes a hanging indent for headings, such that the heading protrudes into the margin, these too can mirror each other, extending into the left margin on even pages and the right margin on odd pages.

Facing pages should also be balanced in length. If one page is filled with text and graphics and the facing page is only a paragraph long, space is wasted and the reader will pause to consider whether material might be missing. A short page usually occurs when large art or a long list follows it. Because the art or list needs to appear all on one page, the preceding text won't fit and gets left behind. Juggle your text, reduce the art, or break the list so the preceding page is full. Ideally, the bottoms of your pages should line up, although most formats allow a little unevenness. The last page of a chapter can be any length, as long as it does not contain a *widow* (the next subsection explains what a widow is).

Begin chapters on an odd-numbered page. If the preceding chapter does not end on an even-numbered page so that the next page is odd, insert a blank page. The blank should be even-numbered and in sequence.

ELIMINATING WIDOWS AND ORPHANS

A *widow* is a short segment of type that gets moved to the next page, and an *orphan* is one that gets left behind. The segment is so short, it seems stranded on the page, cut off from the rest of its meaning as well as visually isolated.

You can consider the last and first lines of paragraphs as widows and orphans if the page break strands them. However, the introduction to a list, no matter what its length, is an orphan if it becomes separated from the list items it introduces. Similarly, any phrase followed by a colon or *em-dash* (a long, introductory dash) should appear on the same page as the material it introduces.

Most word-processing programs now have a "widow control" function that makes sure at least two lines of a paragraph appear on a page. Additionally, some programs have a page-viewing feature that reduces the size of your pages on the computer screen. This feature allows you to see their layout and check facing pages for balance, as well as for widows and orphans. At the very least, current word-processing programs show where pages break, so you can proofread page breaks on your computer screen before printing a draft.

If your word processor does not have these functions, you will have to correct widows and orphans by proofreading a draft of your manual. You will always proofread your final draft to check that introductory elements, like introductions to lists, are not orphaned.

PROOFREADING THE FINAL DRAFT

After you polish your format, you'll perform a number of proofreading checks, some of which have already been described in the chapter, "Editing Your Work." The following production checklist summarizes elements you'll proofread during the production process.

▪ ▪

Checklist 15 – 1.
Production

- Be sure that art (including tables, graphs, and so on) appears after the text introducing it.
- Check that page numbers, running heads, margins, and similar format elements mirror each other between odd and even pages.
- Make sure pages are visually balanced and approximately even in length.
- Check that each chapter begins on an odd-numbered page. Insert blanks where necessary.
- Eliminate widows and orphans.
- Proofread spacing before and after paragraphs, art, and captions to insure consistency for each element.

- Proofread page numbers in cross references, the table of contents, figures list, and tables list.

▪ ▪

INDEXING

For your document to be useful, an index is essential. Don't underestimate its value. Even if your publication department does not usually take the time to index, you can take the time to do it yourself. Think about it this way. Your document going out the door without an index is like the product going out the door without your document.

An index is more than a reference aid. It's the user interface, the guide, the entryway to the information your document contains.

CHARACTERISTICS OF A GOOD INDEX

The reader looks in the index for concepts, operations, and actions, and often does not know the technical term associated with them. In fact, that's what an index is for—to help the uninitiated find the unfamiliar by looking up the familiar. A good index works well for the same reason a good document works well—it anticipates the reader's needs. In fact, the two organizational principles described in the chapter, "Organizing a Writing Project, on Paper and Online," apply to indexes as well. Remember them? They are *recognizable structure* and *task orientation*.

An index has a recognizable structure to anyone who has been to school. The familiar, alphabetical sections need not be explained. However, task orientation for an index means that the index lists terms the reader is most likely to seek. These might be terms that do not even appear in the text.

WHAT AN INDEX IS NOT

An index is not an alphabetical list of words. Unfortunately, that is what you'll get from computer programs, called *index generators,* that automatically create an index. Designed by programmers, these tools generally do not allow for the complex decisions that go into creating a useful index.

Some index generators will automatically list all pages containing a particular term. Many of these pages might offer no useful information about the term. Perhaps it is just used in a sentence. An index generated in this way is practically useless. After the poor reader looks up three or four pages and finds nothing of value, he or she gives up in disgust.

Most index generators let you highlight words in a computer file. The program subsequently lists the words, alphabetically, with page numbers. This is only slightly better than an indiscriminate list. The highlighted list still does not contain most of the words an uninitiated reader will try to find.

The drawbacks of index generators don't render them useless. They excel at finding page numbers associated with terms. You then need to add topics and to edit the index, as described in this chapter.

SELECTING INDEX TOPICS

To select an index topic, first imagine how the reader will approach the book. What information does the reader expect to find and where will he or she look? Then phrase the index entry descriptively, to indicate the kind of information it points to.

Include index entries for:

- Descriptions of important concepts and operations.
- Explanations of actions the reader needs to perform.
- Definitions of terms, abbreviations, and acronyms.
- Restrictions, such as notes and cautions, informing the reader of the product's limitations.

Good index entries are descriptive. They give the reader some sense of what he or she will find on the pages listed. For example, let's say a student pilot wants information about night landings. She looks up *landings* in the index of a flight training manual. She finds the entry lists 29 pages with information about landings. To find information about night landings, or to find out whether or not the book even contains the information, she might have to look on all 29 pages.

A descriptive index entry contains subentries pointing to specific information. The following example shows an entry that would help the student pilot find information on night landings, as well as many other aspects of landing a plane.

LANDINGS
   180° point, 3–10
   aiming point, 3–14, 3–26
   approach, 3–10
   approach airspeed, 3–10, 3–12
   approach angle, 3–13
   at night, 7–21

base leg, 3–11
best angle-of-glide speed, 3–14
bounced, 3–17
crosswind, 3–18

This portion of an index entry from *Private Pilot Maneuvers* tells the reader exactly the kind of information to expect. The entire entry contains 25 subentries, referring the reader to 29 different pages.[1]

FORM OF ENTRIES

Index entries have two parts—a subject and a reference. The reference can be one or more page numbers or a pointer to another subject (*"see* topic"). Certain rules govern the form of index entries.

Entries and Subentries

Index entries can either stand alone or contain subentries. Subentries describe specific aspects of entries. They can be either *nested* (on the next line and indented a few spaces to the right of the main entry) or continuous. The *landings* entry in the preceding section is a good example of a nested entry. A long, detailed index might have more than one level of subentries, with sub-subentries providing even finer detail.

A short, simple index with only one level of subentries can contain entries formatted in a continuous line rather than nested. The following example shows a continuous entry:

searching: multiple text strings, 2–11 through 2–14; single text string, 2–8 through 2–10

See References

A *see* reference appears in place of page numbers and directs the reader to a main entry containing page numbers. Use it for synonyms of the main entry. For example:

ending a session (*see* logging off)

This entry tells the reader to look up *logging off* for pointers to information about ending a session.

A *see* reference contains no page numbers and sends the reader to another part of the index. Therefore, use it only to refer to long entries. If a main entry is small, include page references for the synonym, as well for the main term, and place the main term in parentheses. By including page numbers, rather than using a *see* reference, you save the reader the trouble of looking elsewhere. For example:

ending a session (logging off) 40–41

See Also References

A *see also* reference directs the reader to another index entry for related information. It also expands the reader's search by indicating topics the reader might not have considered looking up. For example, to tell terminal users how to log on to a host computer, you will list *logging on* in the index and include page numbers where logging on is explained. Then, to point the reader to information about how security keeps unauthorized users from logging on, you would include a *"see also* security" reference.

A *see also* reference usually appears as the first subentry, with the words *see also* italicized:

logging on   88–93
    see also security

Page Numbers

Individual page numbers are listed with commas separating them. The following example shows a topic discussed on two separate pages:

fluid levels
    checking   9, 15

Continuing page numbers indicate an ongoing discussion of a particular topic and are listed by first and last number, separated by a hyphen. The following example indicates an ongoing discussion:

fluid levels   9–17

Index generators often don't provide easy ways to indicate continuing page numbers. Therefore, check long lists of individual page numbers in a computer-generated index to see if they are all part of a continuous discussion; then edit the page reference for proper form by listing the first and last page numbers, separated by a hyphen.

CREATING AN INDEX

You can create an index either manually or with the help of an index generator. Because most word processing programs now come with indexing capabilities, you'll probably do the latter. This section assumes you'll use an index generator.

You can use the information in this section to guide you through a manual index as well, but you'll have to list entries and page numbers manually and alphabetize them yourself.

When you index your document, you will

- go through a *hard copy* (computer print-out) of your final draft and mark terms you want to index.
- enter the terms in the index generator.
- run the software to generate the index.
- edit the index.

Marking a Hard Copy

On a hard copy of your finished document, highlight or underline those terms you want to index. Remember to mark only those places where the reader will find significant information about terms; do not mark every occurrence. In the margin next to the highlighted term, write additional entries you'd like to include that refer the reader to the information. For example, if you mark the words *engine maintenance,* you might write in the margin *tune-ups* (see *engine maintenance*).

Also write subentries that describe the kind of information the reader will find on the page. For example, if a page describes how often engine maintenance procedures need to be performed, you might highlight *engine maintenance* and write *frequency* in the margin.

Using an Index Generator

Once you mark a hard copy, you can open your computer file and either highlight the terms online or enter them in whatever way your index generator requires. Because these indexing programs differ quite a bit, details on how to use them are beyond the scope of this discussion.

If you have experience using index generators and the one you're using is particularly flexible, you might be able to highlight terms directly online, without first marking a hard copy. Some index generators allow you to type additional entries and subentries in your document, using *invisible text* (text that does not appear in your printed document but that can be read by the indexing program). These capabilities can save you a lot of time. However, fancy software can provide the illusion that you're creating a good index. Don't skip the step of first learning what makes an index useful. You will always need to edit an index, and one created by computer software needs editing more than one you create without such help.

created by computer software needs editing more than one you create
without such help.

Editing the Index

All indexes, even manually created ones, will contain multiple forms of
the same terms. For example, the entries *logging on* and *log on* might
appear in separate places.

logging on   2–5, 8
log on   8, 17

Because they are synonymous, these entries' page references should
be combined next to the entry the reader is most likely to look up.

logging on   2–5, 8, 17

Once you have cleaned up your topic entries, read through the index
and ask yourself what additional entries and subentries might help someone who is unfamiliar with the product. Add them now.

Next, edit the mechanical errors in your index.

Correct any misspelled terms. If a page reference appears next to a
misspelled term, and the term appears correctly spelled elsewhere in the
index, move the page reference to the correct entry and delete the mis-
spelling.

Delete page numbers that repeat within the same entry. For example,
in the following entry, the first two page references should be deleted:

logging on   2, 3, 2–5

Proofread the index for correct alphabetization.

Proofread page references against the text to make sure that the page
referred to contains the information listed in the index entry. For a long
computer-generated index, you can spot-check page numbers rather than
proofreading every one against the text. If you find even one error, however, your index generator has a bug and you'll need to go back and
proofread all page references.

Proofread all page numbers in a manually created index.

Finally, polish the format of the index as thoroughly as other parts of the document, checking for proper alignment, spacing, and widows and orphans.

▪ ▪

**Checklist 15 – 2.
Editing the Index**

Refer to the following checklist when you edit and proofread an index:

- Combine page references from multiple forms of the same entry (e.g., *log on* and *logging on*) into one entry.
- Create new entries and subentries that will increase the usefulness of the index.
- Delete page numbers that repeat within the same entry.
- Proofread the index for correct alphabetization.
- Proofread page references against the text.
- Proofread spacing, alignment, and other formatting elements.

▪ ▪

SUMMING IT UP

With this chapter, you've come to the end of the documentation process. You now know how to produce a draft that is ready for the printer, complete with index. The next two chapters discuss aspects of the technical writing profession that affect your performance and your future.

▪ ▪

[1] Excerpted from the index of *Private Pilot Maneuvers* (Englewood, Colo.: Jeppesen Sanderson, Inc., 1989), p. I–2.

PART FOUR

DOING BUSINESS

16

THE HAZARDS
OF BEING A
TECH WRITER

"The price one pays for pursuing a profession, or calling, is an intimate knowledge of its ugly side."

James Baldwin (*Nobody Knows My Name*, 1961)

All professions have an ugly part—the part you don't hear about until you're in the middle of it. However, if you know about problems before they happen, you can prepare yourself to deal with them.

Let's look at some of the unpleasant aspects of a high-tech career, such as stress and computer-related injuries, with an eye toward their solutions.

 ## STRESS

Doing business as a technical writer means producing high-quality documentation under often stressful conditions—conditions caused by aggressive deadlines and a variety of irritants inherent in a high-tech environment. Learning to deal with stress can help you avoid burnout and excel at your work.

190

"Job stress costs American businesses at least $200 billion a year—in absenteeism, diminished productivity, direct medical expenses, employee turnover, insurance premiums, and worker's compensation awards," says corporate consultant Robert K. Cooper. "Put in perspective, that's already more than the profits of all the Fortune 500 companies combined."[1]

Every job provides stress. A certain amount of stress creates challenge and makes an otherwise boring job more interesting. Too much stress, coupled with lack of appreciation for your efforts, can lead to burnout.

BEWARE OF BURNOUT

Burnout is "a state of physical, emotional, and mental exhaustion, caused by long-term involvement in situations that are emotionally demanding," say psychologists Ayala Pines and Elliot Aronson.[2]

Tech writers do burn out. Feelings of futility lead to caring less about the quality of a document, becoming cynical about their company and its products, gossiping about uncooperative engineers, and slowing down on the job. How can you avoid burnout? By assessing the stresses in your unique situation and taking control of them before they control you.

STRESSES INHERENT IN TECH WRITING

You already know the most common stresses tech writers face. They're listed in the first chapter of this book. And they're listed again below because now we'll look at them in relation to stress management. They include:

- difficulty obtaining information.
- reticent or uncooperative engineers.
- canceled projects (after the work's been done).
- unreasonable or unclear deadlines.
- unwieldy tools and equipment.
- office politics.

Writers' occupational injuries can be added to this list of stresses. These are the physical problems born of spending long hours using a computer.

STRATEGIES FOR DEALING WITH STRESS

Independent-minded, creative people—the kind of people who become tech writers—tend to blame themselves when they have difficulty coping with a stressful situation. Yet, according to Pines and Elliott, burnout is a situational problem, rather than an individual failing. Most of the stresses that lead to burnout are caused by organizational and environmental problems.

Some stresses can be changed, others cannot. And many of the ones that cannot be changed become bearable if sufficient rewards are present. One of the first strategies for coping with stress is to distinguish between those problems you can solve and those you'll have to live with.

For the stresses that can be changed, take action. The more control you have over setting your own limits, the less likely you are to burn out. To cope with those problems that can't be changed, you'll need a support group of sympathetic peers, some stress-reduction techniques, and the ability to extract either pleasure or meaning from some aspects of your job.

Communicating Problems and Solutions

Most problems that can be changed should be taken up with your manager. It's part of a manager's job to solve problems like unclear or unreasonable deadlines and unwieldy tools. Similarly, problems with office politics can often be "delegated upward." If you recall from the first chapter of this book, writers sometimes are forced to play diplomat between engineering and marketing departments that disagree about how the product is to work. Your manager should be able to get you out of the middle of the more stressful interdepartmental feuds.

Phrase the problem to your manager in a nonblaming way and be ready to suggest one or more solutions. Your manager might not even be aware that the problems exist and might welcome your help solving them.

Get involved in the solutions. For example, volunteer to set up a scheduling procedure for your department's publications and to research the amount of time each phase of manual production takes.

Sometimes a manager won't be there as a buffer, and you'll need to take charge of unrealistic deadlines directly, by confronting a product team. Says Jan Roechel of Apple Computer:

> It's really easy for people to look at the product schedule and say, O.K., your documentation can fit in here, here, and here. That was creating a tremendous amount of pressure for me because, first of all, their schedules probably weren't realistic to begin with, but then I was being pushed to fit their schedule without even looking at what I needed and what the document needed. At that point, I stood up in one of our meetings and said, look, this is the

way I think schedules should be done—I should sit down and look at what's needed for my document. And that's what I did.

Jan discovered what psychologists have discovered about stress. The more you feel in control of the forces affecting your work, the less likely you are to burn out.

Cultivating a Support Group

Stress sometimes creates tension among coworkers, leading to cliquishness, gossip, and mistrust. Avoid judging coworkers. Instead look for opportunities to express your appreciation of them. They're experiencing stress too. And whether problems are solvable or not, you'll benefit from a support group of coworkers who understand the stresses you're experiencing.

Fortunately, the camaraderie of writers is one of the rewards of tech writing. They'll not only understand what you're going through, but can also help with feedback about your work.

Lack of appreciation for your work is a significant factor in burnout. You can compensate for lack of support from an overworked manager by setting up a peer review system. Ask fellow writers for feedback. Offer to give it. Make sure that, in addition to diplomatically phrased, constructive criticism, you include a healthy dose of praise.

Practicing Stress Reduction

Stress takes its toll on the body, in the form of hypertension, heart problems, and other stress-related diseases. Two ways to counteract the physical ravages of stress are relaxation and exercise.

Psychologists recommend taking time to relax after work, through meditation or whatever means works for you. One tech writer I know goes for a long walk with his two dogs. Another practices classical guitar.

Tech writing consultant Carolyn Curtis maintains an elaborate garden of vegetables and native California plants. She advises:

> The natural world is a good antidote to all of this rectilinear, techie kind of stuff. Random vegetation, in particular. If you don't have your own yard, go to some wild place, like the city park or whatever. Just something that doesn't grow in straight lines!

Now imagine doing something that not only reduces stress, but helps keep your weight down, increases your strength, improves your resistance to a long list of diseases, tones and shapes your body, regulates your breathing, helps you sleep better, and increases your sense of well-being. Exercise is one of the few bargains in life.

I used to hate gym class in school and have never been athletic. But my first tech writing job just happened to be across the street from a health club that extended discounts to neighboring businesses. I joined mostly for the jacuzzi. Once I started exercising there regularly, however, I was hooked and have made exercise a habit ever since. I almost never get the winter colds and flu I took for granted before I started exercising. And I notice after a workout a significant reduction in physical and emotional tension.

Don't think you have to join a health club or become a jock. Short, brisk walks, gradually increasing in length, can provide most exercise benefits, at no cost. And if you already enjoy bicycling, swimming, or skiing, do it more often. Start considering it a necessity rather than an indulgence.

Focusing on Rewards

Says biomedical writer Dan Liberthson:

> You've got to be interested in what you're doing to do it well. It was my interest in medicine—which I rediscovered when I went to work as a technical writer for a medical equipment firm—that really led me to make a career as a technical writer. If I hadn't found that interest, I wouldn't have had a career.
>
> I think there are very few people who can be happy if they are not interested in the work that they're doing and in the material that they work with.

If you enjoy the rewards inherent in technical writing—the variety, the creative opportunities, and the constant learning—its everyday stresses will recede into the background of your awareness.

Setting Long-Term Goals

Sometimes a particular work situation is just plain unhealthy. Morale is low, unreasonable pressures are beyond your control, and rewards are few. It's time to decide whether to stay or leave.

One of the greatest stress fighters is a clear long-term goal. If you know where you're going, problems along the way seem much less significant than if they are your whole world.

Pines and Aronson report that, for all the research they've done on causes and cures for burnout, one of its main causes is an individual's failure to find meaning in life.[3] This is one stress that you must change inside yourself, because no amount of situational manipulation can lend your life meaning. Long-term goals can be a deep expression of who you are, encompassing your ideals, whom you wish to become, and what mark you want to make on life.

Perhaps your current, uncomfortable situation is a necessary step toward your long-term goal, and you need to leave honorably, with all your projects successfully completed and your character references pristine. With your eye on the goal, you're constantly reminded of why you are putting up with all the garbage.

Your long-term goal can also help you leave, when the time is right, by providing a vision of a happier future. (The next chapter, "Career Excellence," describes some of the career possibilities open to tech writers.)

WRITERS' OCCUPATIONAL INJURIES

Writers live protected lives, ensconced in temperature-controlled offices, free of health hazards, right? Wrong. As frequent computer users, writers are prone to a host of ailments, which are loosely referred to as Video Operator's Distress Syndrome (VODS). These include musculoskeletal disorders, eyestrain, and possibly miscarriages.

REPETITIVE STRAIN INJURIES

The most common musculoskeletal disorder related to frequent computer use is (again, loosely) referred to as Repetitive Strain Injury (RSI). It's also called cumulative trauma disorder (among other things) and includes tendinitis, carpal tunnel syndrome, and a host of more esoteric-sounding afflictions of the hands, wrists, arms, and shoulders.

According to an article in *The Peninsula Times Tribune,* cumulative trauma disorders make up around half of the occupational illnesses now reported to the Occupational Safety and Health Administration (OSHA). And carpal tunnel syndrome, a disease of the median nerve in the wrist, is "the No. 1 cause for surgery among adults in the United States."[4]

Journalists have been hard hit, with writers on the *Los Angeles Times, Newsday,* and other periodicals seeking medical help for chronic pain. *L. A. Times* reporter Judy Pasternak complained, "It hurt to wash, to eat, to open doors, to pick up a baby."[5] Deborrah M. Wilkinson, also of the *L. A. Times,* reported "constant throbbing pain in my upper arms, neck and shoulder."[6] A significant number of technical writers and other computer operators in the San Francisco Bay area are reporting problems with RSI.

VDT-RELATED INJURIES

From eyestrain to miscarriages, a host of ailments have been blamed on VDTs (video display terminals). "Eyestrain is the single largest category of health complaints among VDT users," according to a 1989 article in *New Directions for Women.*

> In New York, an insurance data processor reports having four eyeglass prescription changes in a year after working on video display terminals (VDTs). A telephone company service representative in Minnesota reports that she is periodically on the verge of blacking out after intensive VDT work.[7]

While many such reports of VODS are episodic—that is, seemingly isolated incidents that don't prove anything—the number of such reports is so startling that many feel more research is needed. As early as 1983, the California-based health maintenance organization, Kaiser Permanente, responded to concerns about VDT emissions by studying 1,583 pregnant women. The results sent chills up the spines of computer manufacturers and dealers. Pregnant women using VDTs over 20 hours a week showed an 80% greater likelihood of miscarrying than women who did not work with computers. Although it did eliminate factors such as age, education, occupation, smoking, alcohol, and drugs as causes, the study did not prove VDTs guilty. However, a more recent Swedish study found that the kinds of magnetic fields coming from VDTs damaged mouse embryos.[8]

What does this mean for tech writers? While opposing factions are debating the hazards, we had best use caution. Fortunately, recommendations for safe computer use are being disseminated by a variety of organizations in the public and private sector.

PREVENTING WRITERS' OCCUPATIONAL INJURIES

Recommendations for preventing computer-related ills include changes to the furniture arrangement and lighting of your work space, improved body positioning during work, and frequent breaks. The following tips can help you create good work habits and a safe work environment.

Tips on Preventing Computer-Related Injuries

- Sit straight, with your lower back supported by an adjustable chair back.

- Both feet should rest comfortably on the floor with thighs parallel to the floor; do not cross your legs.
- Sit at least arm's length from your computer screen.
- Use a copy holder, which you can move from one side of the desk to the other. This will allow you to keep your neck vertical and to vary your head movements, rather than repeatedly looking down at the same spot on the desk.
- Eliminate light sources behind you, which cause screen glare, and in front of you, which compete with the screen. Light should come from the side or above you.
- Use a low-glare screen. If your screen reflects glare, use an anti-glare screen cover.
- When typing at the keyboard, keep your wrists straight.
- Use a detachable keyboard. Keyboards attached to a terminal cannot be adjusted for comfort and safety.
- Lower your keyboard to the level where your shoulders can relax while you type. Some recommend a pillow lap desk or an adjustable keyboard support that clamps to your desk.
- Take a break every hour. Studies have shown that an active break, including stretching and light exercise, is more beneficial in preventing RSI than an inactive one.

Office equipment manufacturers are responding to distressed computer users with specially designed furniture and computer equipment, such as ergonomic keyboards, foot-operated mouse devices, and so forth. These developments should change the work place dramatically over the next few years. Whether they will or not depends on the response of employers to computer-related medical problems. Additionally, the need for much more frequent breaks for computer users may affect the nature of office work in the future. After all, who will be left holding the medical bill?

Hopefully, you won't ever need to know anything about cures for VODS. But you might want to take note of reports indicating acupuncture as a possible alternative to anti-inflammatory drugs or surgery for carpal tunnel syndrome. According to one source, "acupuncture and surgery are both about 80% successful in curing CTS."[9]

If you follow the recommendations in this section, you should turn out prose efficiently and safely. (However, if you are pregnant, you might want to look further into the results of studies on pregnancy and VDT use.)

SUMMING IT UP

This chapter described some of the gloomier aspects of technical writing, such as stress and occupational injuries. But it also gave you some ammunition against them. The next chapter will guide you toward career excellence and provide glimpses of future possibilities.

■ ■

[1] Robert K. Cooper, *The Performance Edge* (Boston: Houghton Mifflin Company, 1991), p. 7.

[2] Ayala Pines and Elliot Aronson, *Career Burnout* (New York: Free Press, 1988), pp. 1–10.

[3] Pines and Aronson, *Burnout*, p. xii.

[4] Ruthann Richter, "Carpal Tunnel Syndrome," *The Peninsula Times Tribune*, July 1, 1990, p. C–6.

[5] Judy Pasternak, "Big Pain in a Computerized World," *This World*, April 2, 1989, p. 15.

[6] Deborrah M. Wilkinson, "Computer Dis-ease," *Essence*, March 1989, p. 14.

[7] Elaine Clift, "VDT's Can Bring on Medical Problems," *New Directions for Women*, July/August 1989, p. 1.

[8] Clift, "VDT's," p. 4.

[9] Gary Perez, "Acupuncture Provides an Alternative," *Connection*, newsletter of the Silicon Valley Chapter of the STC, July 1991, p. 8.

17

■ ■

CAREER
EXCELLENCE

"When men are rightly occupied, their amusement grows out of their work,
as the colour-petals out of a fruitful flower."
John Ruskin, *Sesame and Lilies* (1865)

This chapter describes career development. You'll find suggestions on how to stand out in your current job and how to advance professionally, as well as a detailed section on how to operate as a consultant. You'll also learn about career possibilities open to you as a result of being a tech writer. You can use this chapter to explore your career goals, as a technical writer and beyond.

In the last chapter, you learned about the importance of long-range goals in relation to coping with stress. Long-range goals have the more pragmatic purpose of guiding you in your job choices.

Because tech writers are so individualistic and come from such diverse backgrounds, no single career path is appropriate for everyone. You may want to advance in your corporation, move laterally into a different job category, or strike out on your own. You could also choose to stay a tech writer, changing companies for monetary gain and variety. Or you could choose to get out of the high-tech world entirely. Each choice serves up unique possibilities.

SUCCEEDING AT YOUR CAREER

Writing excellent documentation is usually not enough to promote your career as a technical writer. In the real world, some fairly average writers move ahead because they are punctual, they carry themselves in a professional manner, and mostly because people like them. In the business world, how people see you often weighs more heavily than what you produce.

You can take advantage of this seemingly unfair truth by being punctual, dressing well, and making an effort to get along well with your manager, coworkers, and members of the product team. If you add high-quality documentation to these efforts, you will surely excel.

Your documentation and expertise will go farther if you share them with your professional network. You can enter your work in competitions sponsored annually by the Society for Technical Communication (STC). These competitions provide a wide range of categories, from reference material to technical illustrations. (You'll find the address of the STC and other writers' organizations in Appendix B.)

Once your career is well underway, you may choose to speak at meetings of professional groups. They are always looking for experienced writers to speak on topics like creating a style guide, designing online help, and so on.

By entering competitions and giving talks, you'll become known within your network as a knowledgeable professional. Should you want to change jobs to advance your career, you'll be one of the first in line for choice positions.

CLIMBING THE CORPORATE LADDER

If you are a people person who enjoys tackling problems, consider moving into management. As Susan Tisdale pointed out in the chapter "Why Begin?" being a manager means finding rewards in the elusive stuff of problem solving, human interaction, and performance evaluations. The down side is that you don't experience the tangible results you grew to love as a tech writer—holding that published manual in your hands.

To move into management, begin by taking more responsibility within your department. For example, you can organize a group of interested writers to research and suggest improvements to the company's documentation.

Ask your manager for more responsibility. After you've gained experience as a writer, you can become a mentor or lead writer, teaching newly hired writers about departmental procedures and style guidelines.

Many companies expect senior writers to manage projects undertaken by outside contractors or consultants (described in the next section). If you prove your project management skills, you will eventually earn a supervisory role within your company, or you'll move on to a company where such an opportunity exists.

INDEPENDENT CONSULTING

"We all end up earning a living," says consultant Linda Lininger. "At least with being a free-lance technical writer, I feel that I'm in control of what I do. I earn a living my way."

As an independent consultant, you are a "free-lancer," a "contractor," a self-employed technical writer, and a small business. All these labels will be applied to you at different times, and each carries with it associations and meanings. For example, to the Internal Revenue Service (IRS) you are a small business, but you may be called upon to prove it.

Although I'll use these labels interchangeably, some define them more precisely. In an article in *Technical Communication*, consultants Patricia Caernarven-Smith and Anthony H. Firman make the following distinctions:

> A *contractor* . . . is a person who works at a fixed rate, often for a fixed period, on a defined project. Contractors are hired by contracting houses (also called "job shops") on behalf of a client company.
>
> A *freelancer* is also a person who works at a fixed rate, often for a fixed period, and often on a defined project. Freelancers work directly for the client company, and are hired directly by the client management.
>
> In our business, *consultants* are able to manage projects. Consultants are able to provide information on specific products and the criteria for selecting them. Consultants are able to design publications, determine product liability, help top management plan, and help working management cost and schedule a publication project.[1]

These distinctions are not clear cut but rather part of a continuum. For example, when I struck out on my own, I began as a free-lancer in a company I'd worked for on staff, then contracted through job shops, and finally decided to represent myself without an intermediary. As a self-employed free-lancer, I sometimes perform as a temporary employee and sometimes as a consultant providing advice and services that run the gamut from publisher to political go-between.

The joys of independent consulting include being your own boss and setting your own schedule and conditions. The drawbacks of independent consulting include the constant need to hustle work (which is not a drawback if you like hustling), to keep accurate financial records, to know tax laws, and to assume entrepreneurial risk. I and many others have chosen the independent route because of the freedom it affords.

BECOMING A CONSULTANT

Getting started as a consultant really means getting started as an independent business person. The following prerequisites will help you succeed:

- business contacts
- a good track record
- knowledge of publishing procedures
- ability to budget for lean times
- willingness to assume entrepreneurial risk
- ability to sell yourself
- skill at negotiating agreements
- aptitude for accounting and for keeping on top of changing tax laws

You can see that this is not a small undertaking. But a time may come when you'll have the experience and contacts you need to generate a lucrative consulting business. And you can begin without knowing all there is about taxes and such. You can learn as you go along, and others who've gone before can help with the details.

Your success hinges a great deal on your contacts—who you know. That's one of several good reasons to work in the field for awhile before you go it alone. While you build your career as a staff writer, you establish a reputation with managers, supervisors, and lead writers. This reputation will follow you.

For example, six technical writers I've worked with on staff have since become supervisors or managers. Five of them have provided me with contract offers, either directly or through a recommendation to someone else in their company.

Another good reason to work in the field for awhile before becoming a consultant is that you will need excellent writing samples, references, and the ability to "land on your feet." As a consultant, you will be expected to know industry-standard procedures for producing whatever kind of documentation is your specialty. You will need to be familiar with how a technical publications department works. Or if you plan to free-lance for technical periodicals, corporate newsletters, or some other publishing environment, you will need to know their production and scheduling processes.

If money burns a hole in your pocket, don't become a free-lance writer. That is, unless you're comfortable living "close to the bone." To use yet another cliche consultants use frequently—it's either feast or famine.

"I was really nervous in the beginning of my contract career because it was a lousy year to get started," says consultant Carolyn Curtis. "It was 1985, which was a terrible year in the industry. If I hadn't had a cushion saved up, I would have really been in trouble."

Says consultant Cal Callahan in an article in *Technical Communications:*

> My idea of a prudent reserve is enough money to call a tow truck, so I'm often pennies away from poverty while a $5000 manual is being reviewed. Not too long ago I went out and upgraded my CP/M computer to a Big Blue compatible, forking over about $1000 in cash. Then a $2200 contract I had anticipated was delayed for two weeks. Was I low! Then that contract came through. Along with it came the go-ahead on another big job from the same client, plus major changes to another manual from another client, plus two new brochures. Whooppee, here we go again![2]

That's luck! During my first years consulting, I turned down a couple of contracts—one because it seemed tedious, another because I sensed conflicting expectations within the organization. Then I couldn't find another contract, and the phone didn't ring for several months. I spent long hours questioning my wisdom in turning down those jobs.

Some consultants advise taking whatever opportunity comes along. Others advise selecting only those at which you know you'll excel and be happy. I still ascribe mostly to the latter point of view. You'll have to find the level of income and job satisfaction that meets your particular needs.

MEETING CLIENTS' EXPECTATIONS

As a consultant, you are often called in to "fight fires"—to handle problem projects encumbered by difficult personalities, an unrealistic deadline, or both. Usually these projects are considered undesirable by staff writers, and you are seen as the magician who'll be able to handle them.

The positive way to look at this is that you can handle them. You are not embroiled in whatever political battle is raging around a particular project, you are seen as an outside expert—a possible bearer of solutions and sanity, and you can set conditions under which you will perform. For example, you can say you'll meet the (unmeetable) deadline, provided that the product is finished and all information is in your hands by a certain date. You can establish contingencies—If the product isn't finished by that date, the deadline will slip.

As a consultant, you'll be expected to ask informed questions about your client's company procedures. For example, when is beta test? Has a product freeze date been set? Will I be responsible for camera-ready copy?

You will also be expected to produce at a faster pace than a staff writer would.

Some clients might expect a periodic status report, detailing how you've spent your time. It's a good idea to provide at least a general description of your activities when you submit your invoice. Linda Lininger goes even farther:

> I keep status reports weekly. In my daily calendar, I make notes about what I do with my time, so that when I do my status report I can write. "I spent this many hours in my day at the keyboard and I edited these files. I spent this much time formatting and printing, and these are the files, or this is what I did." So that when I write a status report, I can account for how I wish to charge my client.

The most important thing to remember is that each client's needs are unique. Try to sense what the client expects from you. What can you provide that will be perceived as meeting their needs? Is it a perfectly formatted document? Will a crude, meticulously accurate one do? It may be more important for you to attend meetings and communicate on behalf of the publications department than to produce anything.

If after talking with a client about their expectations, you are still not clear about how to meet them, ask specifically "What would you like to see as evidence that I am meeting your needs?"

DOING BUSINESS AS A CONSULTANT

Once you decide to free-lance, you can begin to learn about business practices. Should you get a business license or incorporate? Should you engage in "fixed-bid" contracts or charge an hourly rate? What contract terms or agreements will protect you?

The answers to these questions fill books and are constantly changing. For example, the tax implications of incorporating and independent consulting change almost annually. Engage an accountant who is knowledgeable about self-employment to help you with the details.

Licenses, Incorporating, and the IRS

You can do business as a technical writing consultant without any special licenses. However, occasionally the IRS might require proof that you are independent, because they want to be sure your client is not getting out of paying the taxes they would have to pay for you as a regular employee. The definition of a consultant is somewhat vague in the eyes of the IRS, but things they look for are "assumption of risk," a work place off site, multiple clients (as opposed to one client who supplies all your income), and so on. Having a business license or being incorporated is another sign that you are independent. The best way to keep on top of this issue is to join an organization like the STC or the National Writers Union (see Appendix B for addresses) and to network with fellow consultants.

In general, incorporating is a hassle, forcing you to engage an accountant on a regular basis and in some states (like California) requiring a large minimum tax payment. Some consultants feel it's worth is because it shelters their personal assets from liability and, with the help of a particularly agile accountant, can sometimes save on taxes. Most tech writing consultants don't bother with incorporating.

Written Agreements

You will need a written agreement of some sort between you and your client. Larger companies provide a set contract. Read it before you sign it. If it contains the words "work for hire," you are signing away the copyright to your work. That's O.K. in consulting, and after all, where else are you going to sell a manual on electronic mouse traps except through the company that makes them? However, most companies are not aware of copyright law and so do not include those specific words in the contract. If "work for hire" does not appear in your contract, you own the copyright, and this little detail can help you a lot—if the client claims they cannot pay you, for whatever reason, you can prevent them from publishing your manual.

Even if you sign a company-provided contract, you should write a letter of agreement and ask your client to sign it. The letter should reiterate any verbal agreements you have made with the client, such as your rate of pay, the name of the product you're documenting, the kind of document you've been asked to produce, the time frame, and so on.

While verbal contracts are legally binding, they're awfully hard to prove without a paper trail.

A few tech writing contractors prefer to work on a *fixed-bid* contract. *Fixed-bid* means you will be paid a lump sum when you meet one or more milestones, which usually correlate with document drafts. For example, you'll be paid 50% of the sum for a first draft, 25% for a second review draft, and the final 25% for camera-ready copy. This kind of agreement requires you to produce a fairly elaborate written contract, detailing, among other things, contingencies that might inflate your fee. A fixed-bid forces you to estimate your hours very accurately, so that you don't wind up working for free. And you can never allow for all the unforeseen problems that arise during a product release.

Most consultants I know shy away from fixed-bid contracts, because contingencies are inevitable and such contracts often evolve into hourly arrangements anyway.

Overhead and Setting Rates

As a consultant, you are expected to own your own equipment. A computer and printer are the minimum you'll need to begin. A fax machine, modem, and miscellaneous items like a paper cutter are also helpful.

Remember that your equipment and office space in your home are overhead and should be taken into account in setting your rates. Figuring rates is too involved to discuss thoroughly here. The Independent Consultants' Special Interest Group of the STC can provide information on rate setting, as can books on consulting.

WORKING THROUGH JOB SHOPS

Job shops, job brokers, "jobbers," recruiters, contracting houses—all these are names for the agencies that will represent you as a contractor to clients, cut your check, and skim a hefty percentage off the top. Their cut can range from 30 to 50% of cost billed the client. This means they might bill the client $55 an hour, and you will see around $35 an hour in pay.

Despite the excessive amount they skim, job shops can help you get started as a free-lancer. They provide the contacts, and you provide the rest—your writing samples, references, skills at being interviewed, and finally your skills on the job will establish your value and build contacts in the free-lance market.

Before you apply to a job shop, check its reputation carefully. A startling number of job shops in Silicon Valley are known for misrepresenting the nature of assignments, for misrepresenting writers' skills, and worst of all, for not paying writers. The reputable job shops are reasonably honest about their clients' expectations and will pay you on time, even if the client is late.

In response to excessive job-shop fees, the National Writers Union (NWU), an organization that promotes fair conditions for all kinds of writers, sponsors a tech writers' hotline. The hotline currently serves the San Francisco Bay area, but has plans to expand nationally. This service brings together contract tech writers and clients at a greatly reduced fee. To use the hotline, you would need to join the NWU (see Appendix B for the address). On a national basis, the NWU also provides a grievance committee of negotiators who'll help you deal with clients and job shops that don't pay you.

 ## WHAT'S NEXT?

The average American worker changes jobs every 3.6 years and will work for 10 different employers in a life time, report Hizer and Rosenberg in *The Resume Handbook*. "In addition, four out of five job-hunters seek to *change careers* at least once."[3]

Even though tech writing is a rewarding, creative endeavor, you may choose to move on at some point. Besides the climb toward management or consulting, other career paths become more accessible as a result of your years of technical writing experience.

LATERAL CAREER CHANGE

Perhaps technical writing has sparked your interest in technology. With your product knowledge, you can aim for a technical support job within your company, changing careers without having to pound the pavement or even return to school. Many technical writers can handle customer support calls with a minimum of extra training. Providing field support at customer sites is another option for technically knowledgeable writers willing to get their hands dirty.

Training is another obvious choice for extroverted tech writers who can speak knowledgeably about a product. Instead of writing manuals and training material, you'll verbally communicate material written by others to a classroom full of tech support trainees, new employees, or customers.

For the yet-more-extroverted, opportunities exist to move into marketing and sales.

Within the biomedical field, you can aim for a clinical research associate position, monitoring the outside agencies that perform drug trials. As a clinical research associate, you might design how the study is to be done and act as a liaison between the pharmaceutical company and the testing firm.

And as in other technical fields, your options as a biomedical writer also include training, education, marketing, and sales. In training and education, you can provide project management, supervising outside agencies and contractors in writing training materials. In marketing, you devise a marketing strategy for a specific drug, plan its advertising campaign, and contract with ad agencies to implement the campaign. Additional opportunities exist in public relations, arranging speakers' tours for noted physicians to address groups of fellow physicians on new developments in biomedical research.

OTHER KINDS OF WRITING

Many technical writers have other writing interests. Some write children's books, others science fiction, others poetry. Just yesterday, a fellow free-lancer told me she's writing books for disabled children, explaining their disabilities to them. My avocation is travel writing.

Technical writing prepares you for many of the demands all writers experience—churning out prose to deadline, accepting editing, and dealing with criticism. While most writing pays less than tech writing, it may be your next career choice.

Says Dirk van Nouhuys, veteran of 29 years of tech writing and published fiction writer:

> Many people get into technical writing who write fiction or poetry, as do I, and they say how does it affect your writing? And I have an answer for that which is at first it's a benefit, because it's a benefit to your writing to learn any disciplined subset of writing.
>
> So you learn how to do technical writing, which characteristically has kind of a short breath. It has short sentences and small organizational units. It's very modular. And it's a benefit to learn any writing discipline as a writer. After a certain point—after a certain number of years—that becomes a problem because that kind of gets into your blood. And that's not the only way to write. And there are a great many things that you don't write about in technical writing, like birth, death, love, hate, terror, anxiety. [Laughs] You don't write about them. And your fingers can kind of lose touch with writing about them and with those kind of words—truth and beauty—and so after awhile you have to be able to remove yourself from the technical writing mind set to write other kinds of writing.

Whatever you choose to do next, be it managing, consulting, fixing modems, or writing the great American novel, technical writing will give you many of the skills you'll need.

 ## SUMMING IT UP

This book described to you the joys and pitfalls of technical writing. It allowed you to see into the lives of working writers and the minds of hiring managers. It guided you through the steps of producing a technical document, from scheduling and research right through to last-minute formatting of a camera-ready copy.

Then, this book went on to describe the worst and best of doing business as a tech writer. This final chapter looked at career development and long-term goals.

In all, you are now more ready to pursue a technical writing career than if you had not read this book. You can always refer to it for guidance as you go along. I wish you the best of luck.

▪ ▪

[1] Patricia Caernarven-Smith and Anthony H. Firman, "Consultant, Freelancer, Contractor: Take Your Choice," *Technical Communication,* Journal of the Society for Technical Communication, Fourth Quarter 1986, pp. 218–222.

[2] J. F. Callahan, "A Beginner's Guide to Freelancing," *Technical Communication,* Journal of the Society for Technical Communication, Fourth Quarter 1986, p. 212.

[3] David V. Hizer and Arthur D. Rosenberg, *The Resume Handbook* (Berkeley, Calif.: Ten Speed Press, 1977), p. 19.

APPENDIXES

APPENDIX A:
COLLEGES WITH
INTERNSHIP
PROGRAMS IN
TECHNICAL
WRITING

The table[1] in this appendix describes some of the graduate and undergraduate internship opportunities that have been offered through college and university technical communication programs in the past.

Most of these programs are still operating and more programs begin each year. Write to the schools that interest you for details about their offerings.

■ ■

[1] Reprinted with permission from *Technical Communication*, Journal of the Society for Technical Communication, Third Quarter 1990.

Table A-1 Summary of College and University Internship Programs as of June 27, 1989

States and Schools	Students Willing to Travel	Level of Program Grad (G) Undergrad (UG)		Paid?	Length of Internship	Work Hrs/Week
ALABAMA						
Univ. of Alabama in Huntsville	none	UG only		optional	1 quarter (10-wk)	15
ARKANSAS						
Univ. of Arkansas, Little Rock	few	G and UG		optional (G: usually paid)	UG: 1 15-wk term G: 300-hr min	UG: under 20 G: 20–40
CALIFORNIA						
California State Univ., Long Beach	NA (not known)	G and UG		optional (usually paid)	45 hrs/unit (up to 2 units)	varies
San Diego State Univ.	some (if paid position)	G and UG		some do	1 semester (15-wk)	under 20
COLORADO						
Colorado State Univ.	some	G and UG		UG: optional G: preferred	1 semester* (14-wk)	5–40
Metropolitan State College	few	UG only		optional	1 semester* (15-wk; 10-wk summer)	20–40
Univ. of Colorado in Denver**	some	G and UG		optional	G: 1 semester *UG: 1 semester (16-wk or 8-wk)	G: 20–40 UG: up to 40
DELAWARE						
Univ. of Delaware	some	UG only		optional	1 semester* (14-wk)	under 20

Table A-1 Summary of College and University Internship Programs as of June 27, 1989 (Continued)

States and Schools	Students Willing to Travel	Level of Program Grad (G) / Undergrad (UG)	Paid?	Length of Internship	Work Hrs/Week
FLORIDA					
Florida Institute of Technology	few	UG only	optional (usually paid)	1 quarter (10-wk)	20–40
Univ. of Central Florida	most	G and UG	yes	1 16-wk term or 12-wk summer	40
GEORGIA					
Southern College of Tech. (internship program to begin fall 1989)	don't know	G only	yes (usually)	1 or more terms	30 (1 term) 15 (2 terms)
ILLINOIS					
Illinois Institute of Technology	G: some UG: few	G and UG	yes	1 term (10–15 wk)	about 20
IOWA					
Iowa State Univ.	some	G only	optional (seldom paid)	1 semester* (14-wk)	up to 20
KANSAS (no respondents)					
LOUISIANA					
Louisiana Tech Univ.	most	UG only	optional (usually paid)	1 quarter (10-wk)	20–40
Northeast Louisiana Univ.	few	UG only	optional (usually paid)	1 semester or 1 summer (14-wk or 8-wk)	20–40
MAINE					
Univ. of Maine	few	UG only	optional	1 semester* (14-wk)	up to 20

Institution	Extent	Pay	Level	Duration	Hours/wk
MASSACHUSETTS					
Northeastern Univ.	some	yes	G only	1 quarter* (12-wk)	20–40
MICHIGAN					
Michigan Technological Univ.	some	optional	G and UG	negotiable G: 1 only UG 1*	20–40
MINNESOTA					
Mankato State Univ.	G: some UG: few/some	optional (usually paid)	G and UG	1 term (10-wk)	G: 20–40 UG: under 20
Univ. of Minnesota	some	yes	G and UG	1 quarter	20–40
MISSISSIPPI (no respondents)					
NEBRASKA					
Univ. Nebraska Medical Center	NA (part-time)	no	G only	30 weeks	20–40
NEW MEXICO					
New Mexico State Univ.	some	optional (usually paid)	G only	1–2 terms	1 term: 40 2 terms: 20
New Mexico Tech	most	optional (usually paid)	UG only	1 semester* (usually summer) (16-wk or 8-10 wk)	20–40
Univ. of New Mexico	few	optional (usually paid)	G and UG	1 semester* (16-wk)	under 20
NEW YORK					
Clarkson Univ.	few	no	UG only	1 semester* (14-wk)	up to 20
New York Institute of Technology	few (unless reimbursed)	optional (usually paid)	UG only	1 semester (15-wk; 10-wk summer)	up to 20 (8 required)
Rensselaer Polytechnic Inst.	few	optional	G and UG	1 semester* (16-wk)	10

**Table A-1 Summary of College and University Internship Programs
as of June 27, 1989** (Continued)

States and Schools	Students Willing to Travel	Level of Program Grad (G)	Undergrad (UG)	Paid?	Length of Internship	Work Hrs/Week
NORTH CAROLINA						
East Carolina Univ.**	some	G and UG		optional	1 semester (14-wk)	20–40
North Carolina State	G: some UG: few	G and UG		G: yes UG: no	1 semester (14-wk)	G: 20–40 UG: under 20
OHIO						
Bowling Green State Univ.	most	G and UG		yes	1 semester* (14-wk)	20–40
Miami Univ.	some	G and UG		G: yes UG: optional	1 semester	G: full-time UG: varies
Youngstown State Univ.	some		UG only	optional (usually paid)	1 quarter (10-wk; 5-wk summer)	10 and up
OKLAHOMA						
Oklahoma State Univ.	G: some UG: few	G and UG		optional	1 semester or 1 summer (16-wk or 8-wk)	8 (120/sem)
OREGON						
Oregon State Univ.	most	G and UG		optional (usually paid)	30 hrs = 1 credit	20–40
Univ. of Portland	some	G and UG		no	1 semester	up to 20
PENNSYLVANIA						
Carnegie-Mellon Univ.	G: most UG: few	G and UG		G: yes UG: optional	G: 1 summer UG: 1 semester* (12-wk; 14-wk)	G: 20–40 UG: up to 20
Drexel Univ.**	few		G only	yes	2 quarters (6 months)	20–40

Institution					
Penn State	G: few / UG: some	G and UG	optional	1 term (15-wk)	under 20
Univ. of Pittsburgh	none	UG only	occasionally	1 semester*	up to 20
West Chester Univ.	few	UG only	no	1 semester (15-wk)	under 20
TENNESSEE					
Memphis State Univ.	few	G and UG	yes	1 term	20–40
Tennessee Tech Univ.	some (unless paid)	UG only	usually	1 term (10-wk)	up to 20
Univ. of Tennessee in Chattanooga	few	G and UG	optional (seldom paid)	1 semester (15-wk)	up to 20
Univ. of Tennessee, Knoxville	few	G and UG	yes	1 semester (15-wk)	G: 20–40 / UG: under 20
TEXAS					
North Texas State Univ.**	few	G and UG	yes	open	20–40
Texas A&M Univ.	all	G and UG	yes	1 semester + 1 summer	20–40
Univ. Texas San Antonio	few	UG only	no	1 semester (15-wk)	6
UTAH (no respondents)					
WASHINGTON					
Eastern Washington Univ.	G: some / UG: few	G and UG	optional	1 quarter* (10-wk)	G20–40 / UG: up to 40
Univ. of Washington	some	G and UG	optional (usually paid)	1 quarter (10-wk)	G: 20–40 / UG: 12–40
WEST VIRGINIA					
Alderson-Broaddus College	most	UG only	optional	1 semester (15-wk)	up to 20
WISCONSIN					
Univ. of Wisconsin, Stout**	most	UG only	yes	1+ terms	20–40

* More than one internship experience permitted.
** All interns must be in Co-op program.

APPENDIX B:
PROFESSIONAL
ASSOCIATIONS

This appendix contains a partial list of associations that promote technical writing as a profession or further some aspect of technical writing. The list is partial because new groups appear often. Some of the addresses change as officers change, but the addresses below should give you a start in contacting the associations that will meet your networking needs.

American Medical Writer's Association (AMWA)
9650 Rockville Pike
Bethesda, MD 20814

Special Interest Group for Documentation (SIGDOC)
Association for Computing Machinery (ACM)
11 West 42nd Street
New York, NY 10036

Association for Business Communication
University of Illinois
608 South Wright Street
Urbana, IL 61801

Association of Petroleum Writers
2659 South Quebec
Tulsa, OK 74114

Association of Teachers of Technical Writing
c/o Dr. Carolyn D. Rude
Department of English
Box 4530
Texas Technical University
Lubbock, TX 79409

The Australian Communication Association (ACA)
Department of Communication and General Studies
Queensland Institute of Technology
GPO Box 2434
Brisbane, Australia 4001

Aviation/Space Writer's Association
17 S. High Street, Suite 1200
Columbus, OH 43215

Computer Professionals for Social Responsibility (CPSR)
P.O. Box 717
Palo Alto, CA 94301

Council of Biology Editors
Department of Anatomy
University of New Mexico
Albuquerque, NM 87131

IEEE Professional Communication Society
IEEE (Institute of Electrical and Electronic Engineers)
345 East 42nd Street
New York, NY 10017

Independent Computer Consultants
Box 27412
St. Louis, MO 63141

International Association of Business Communicators
870 Market Street, Suite 940
San Francisco, CA 94102

National Association of Government Communicators
7204 Clarendon Road
Washington, DC 20014

National Association of Science Writers
P.O. Box 294
Greenlawn, NY 11740

National Writers Union
837 Broadway, #203
New York, NY 10003

Professional and Technical Consultants Association (PATCA)
1330 South Bascom Ave., Suite D
San Jose, CA 95128

Society for Technical Communication (STC)
815 15th St. N.W.
Washington, DC 20005

■ ■

APPENDIX C:
DOCUMENT PLAN

This appendix contains a sample document plan. You'll use a "doc plan" to help organize your writing project as well as to present your ideas for review and approval by other members of the product team. Once team members "buy off" on the plan, they are less likely to suggest sweeping organizational changes later, in the review process.

You can use this document plan as a starting point for designing your own. As you send out and get back whatever form you create, you'll modify it to better fit your publication procedures.

April 16, 1992

TO:
Susan Black, Engineering   Harold Schwartz,
Grant Smith, Engineering   Engineering
Lisa Downing, Marketing   Jane Townsend, Marketing
Betsy Long, Technical
    Publications

FYI:
George Freeman, QA   John Ferrero, Technical
Pete McMann, Customer   Publications
    Support   Jim Hartford, Training
Pam Brown, Technical
    Publications

FROM: Janet Van Wicklen

RE: SuperINDEX User's Guide Document Plan

Please review the attached document plan for organizational clarity and completeness. You have received
this because you are on my list of reviewers for the
document. If you will be unable to review the document
during the review periods shown in the document schedule, please assign someone else in your department and
let me know whom you've assigned.

Your signature below indicates your approval of the
document plan. Please return it to me with your comments no later than April 23, 1992.

_____

I approve the document as is.   ☐

I approve the document with changes as marked.   ☐

Name   Department

_____

Document Plan for
SuperINDEX User's Guide
Software Version 1.0
April 16, 1992

Writer: Janet Van Wicklen

Product purpose: SuperINDEX allows users to index CAD drawings and to retrieve them by typing a keyword.

Audience: Readers are engineers and technicians who are thoroughly familiar with their own CAD software or who have access to CAD documentation. Therefore, this manual will not contain guidance on using CAD features.

Content: The objective of this user's guide is to provide step–by–step instructions for all SuperINDEX operations. For details, please refer to the attached outline.

Physical
characteristics: The manual pages will be 5 1/2'' × 8 1/2'', 3–hole punched, and placed in a binder with the software diskette. Each chapter will be marked by a color tab.

SuperINDEX Document Plan   page 2 of 3

SuperINDEX User's Guide
–––––––––––
OUTLINE

Chapter 1. Introduction
           This chapter will begin with a quick–reference
           table telling users how to find information.
    I.   Overview of the product's operations
    II.   Overview of SuperINDEX utilities
    III.   Software and hardware requirements
Chapter 2. Using SuperINDEX
    I.   Starting SuperINDEX
           [Figure 2–1 SuperINDEX Option Buttons]
    II.   Selecting Option Buttons
    III.   Using the StartINDEX Utility
           [Figure 2–2 SuperINDEX Main Menu Option Buttons]
    IV.   Using the Browse Utility
           [Figure 2–3 Browse Menu]
           [Figure 2–4 Text Viewer]
    V.   Using the Smart Index Utility
           [Figure 2–5 Example of Smart Index Search]
    VI.   Executing a Text Search
           [Figure 2–6 Text Search Menu]
           A. Boolean Search Syntax
           B. Query Results
Chapter 3. Creating an Index
    I.   Before You Index
    II.   Word Index
           [Figure 3–1 Word Index Menu]
           A. Word Index Function Keys
           [Figure 3–2 Select All Window]
           B. Word Index Options
           [Figure 3–3 Word Index Options Menu]
    III.   Build Index
Chapter 4. Installing SuperINDEX
    I.   Running Install
    II.   Running Setup

SuperINDEX Document Plan   page 3 of 3

Documentation Schedule
for the
SuperINDEX User's Guide

Mar 22–Apr 14   Research (3 weeks)
Apr 14–16   Write doc plan (2 days)
Apr 16–23   Doc plan review
Apr 24–26   Revise doc plan (2 days)
Apr 26–Jun 1   Write 1st draft (5 weeks)
Jun 1–8   First review
Jun 8   Review meeting
Jun 8–22   Incorporate changes (2 weeks)
Jun 22–29   Second review
Jun 30–Jul 14   Complete final draft; incorporate final
                art (2 weeks)
Jul 14–28   Final indexing and production (2 weeks)
Jul 29–Aug 31   Printer turn–around (4 1/2 weeks)
Sep 1   Manuals in stock for distribution

ANNOTATED BIBLIOGRAPHY

This bibliography is organized by topic. The following topics appear in alphabetical order:

- Careers and job hunting
- Desktop publishing
- Education
- History of technical writing
- Interviewing skills
- Writing

Within each topic, books are listed alphabetically by author.

CAREERS AND JOB HUNTING

Berman, Eleanor. *Re-Entering: Successful Back-to-Work Strategies for Women Seeking a Fresh Start*. New York: Crown Publishers, Inc., 1980.

This book provides very practical skills assessment tools for women re-entering the job market. These tools are equally applicable for career changers of either sex. The book explores the value of going back to school and provides guidance on resume preparation, getting a foot in the door, and performing well in an interview.

Bestor, Dorothy K. *Aside from Teaching, What in the World Can You Do? Career Strategies for Liberal Arts Graduates.* Seattle: University of Washington Press, 1982.

Chapter 12, "Writing, Research, and Other Opportunities in Business," contains a writing and editing section on technical writing.

Bolles, Richard Nelson. *What Color Is Your Parachute? A Practical Manual for Job-Hunters & Career-Changers.* Berkeley, Calif.: Ten Speed Press, 1992.

Updated annually, this classic guide on choosing a career helps you figure out what you do best, and more importantly, what you best enjoy doing. It guides you toward a successful and satisfying career choice; then gives you tips on getting a job within your chosen field.

Career Associates. *Career Choices for Students of English.* New York: Walker and Company, 1985.

The section on tech writing offers a description of the profession, professional and personal qualifications, and geographic information about where most tech writing jobs can be found.

Cooper, Robert K. *The Performance Edge.* Boston: Houghton Mifflin Co., 1991.

In this book, consultant Cooper gives strategies for handling stress and increasing effectiveness on the job.

Gale, Barry, and Linda Gale. *Discover Your High-Tech Talents: The National Career Aptitude System and High-Tech Career Directory.* New York: Simon and Schuster, 1984.

The authors claim that you can test your suitability for a job in high-tech industries by testing your logical, mechanical, and numerical skills. They provide three such tests, tell you how to score them, and go on to describe high-tech careers and jobs.

Hizer, David V., and Arthur D. Rosenberg. *The Resume Handbook.* Boston: Bob Adams, Inc., 1985.

This handbook provides 25 examples of good resumes and five examples of bad ones, with terse pointers to their good and bad qualities. This is an excellent source of organizational and formatting ideas for resumes.

Lathrop, Richard. *Who's Hiring Who*. Berkeley, Calif.: Ten Speed Press, 1977.

This book provides ways to explore your job-related skills; then present them attractively—both on paper and in the interview—to prospective employers.

Medley, H. Anthony. *Sweaty Palms: The Neglected Art of Being Interviewed*. Berkeley, Calif.: Ten Speed Press, 1984.

Medley gives thorough advice on how to win at a job interview, including how to prepare, communicate, reduce stress, dress, and even negotiate salary.

O'Brien, Mark. *High-Tech Jobs for Non-Tech Grads*. Englewood Cliffs, N.J.: Prentice-Hall, Inc., 1986.

Written by a high-tech securities analyst, this book offers advice on how to pick an employer (and a technology) that's not about to go under financially.

Perlmutter Bloch, Deborah. *How to Write a Winning Resume*. Lincolnwood, Ill: VGM Career Horizons, 1986.

This workbook provides worksheets and questionnaires to help you recall and describe the experiences you'll include on your resume. It helps you phrase and format the information attractively and provides examples of different types of resumes. This book is useful if you've never written a resume and want an in-depth introduction to the process.

Pines, Ayala, and Elliot Aronson. *Career Burnout: Causes & Cures*. New York: Free Press, 1988.

Two psychologists explore the situational and attitudinal forces leading to job burnout, and they provide coping strategies.

Winkler, Connie. *Careers in High Tech*. New York: Arco Books, Prentice-Hall Press, 1987.

The chapter devoted to technical writing provides ideas for job opportunities, describes what tech writers do, and describes different kinds of technical publications.

DESKTOP PUBLISHING

Makuta, Daniel J., and William F. Lawrence. *The Complete Desktop Publisher*. Greensboro, N.C.: COMPUTE! Publications, Inc., 1986.

This book provides a somewhat technical look at the history, mechanics, and aesthetics of desktop publishing. A good resource for the computer-literate reader.

Price, Jonathan, and Carlene Schnabel. *Desktop Publishing*. New York: Ballantine Books, 1987.

Price and Schnabel provide highly readable explanations of the elements of page design, from font styles to layouts, with lots of examples.

EDUCATION

Kelley, Patrick M., et al. *Academic Programs in Technical Communication* Washington, D.C.: Society for Technical Communication, 1985.

This book lists and describes 56 academic programs in technical communication. It provides a summary of degree requirements and the names and addresses of the people in charge of the programs.

HISTORY OF TECHNICAL WRITING

Allbutt, Sir T. Clifford. *Notes on the Composition of Scientific Papers*. 3rd ed. London: Macmillan and Co., Ltd., 1923 (originally published in 1904).

This is probably the first book on technical writing. After trudging through piles of presumably hideous stuff, Sir Clifford felt compelled to offer his advice to his Cambridge medical students on how to write. His advice on brevity, and other writerly virtues, is charmingly couched in ornate Victorian English.

Alred, Gerald J., et al. *Business and Technical Writing: An Annotated Bibliography of Books, 1880–1980*. Metuchen, N.J.: Scarecrow Press, Inc., 1981.

An excellent source for tracing the development of technical writing as a recognized profession by tracing the evolution of books about the subject.

Drachmann, A. G. *The Mechanical Technology of Greek and Roman Antiquity: A Study of the Literary Sources.* Madison: University of Wisconsin Press, 1963.

Drachmann describes the work of the world's first known technical writers—Hero of Alexandria, Strato, and others—who described the first pulleys, screws, cog wheels, and other inventions of the Greeks and Romans.

Rink, Evald. *Technical Americana: A Checklist of Technical Publications Printed Before 1831.* Millwood, N.Y.: Kraus International Publications, 1981.

Rink provides an annotated bibliography of the earliest handbooks, monographs, and other publications in the United States addressing technical topics. Technical topics covered the spectrum from how "to colour hats green on one side" to "descriptions of any curious machinery and improvements in arts and sciences." This is a fun bit of Americana! Rink also lists the libraries that hold these historic documents.

INTERVIEWING SKILLS

Farber, Barry. *Making People Talk.* New York: William Morrow and Company, Inc., 1987.

New York radio talk-show host Barry Farber provides a richly anecdotal, humorous pep talk on the art of bringing out people's best. His advice can be applied to interviewing, job hunting, or any situation where you need to positively impress someone.

Metzler, Ken. *Creative Interviewing: the Writer's Guide to Gathering Information by Asking Questions.* Englewood Cliffs, N.J.: Prentice-Hall, Inc., 1977.

Full of anecdotes, this is a very readable, entertaining, and informative treatment of journalistic interviewing. Metzler goes into all stages of the interview process—background research, stages of the interview, categories of questions, and more.

Sherwood, Hugh C. *The Journalistic Interview*. New York: Harper & Row, 1969.

This discussion of the journalistic interview applies to tech writers as well as journalists. Sherwood covers the whole process, from preparing for the interview through bringing it to a close.

Stano, Michael E., and N. L. Reinsch, Jr. *Communication in Interviews*. Englewood Cliffs, N.J.: Prentice-Hall, 1982.

This book provides a detailed analysis of the interview as a communication process. It describes the different stages of the interview, classifies interview questions ("vocalizations") and interview structures, and applies these classifications to different kinds of interviews (for example, journalistic). Good food for thought.

▌ WRITING

Bates, Jefferson D. *Writing With Precision*. Washington, D.C.: Acropolis Books Ltd., 1978.

Bates uses a lively tone with lots of examples to describe principles of expository writing and editing.

Bell, Paula. *HighTech Writing: How to Write for the Electronics Industry*. New York: Wiley Interscience, 1985.

Among other things, this book provides a detailed discussion of the different parts of a technical document.

Brockmann, R. John. *Writing Better Computer User Documentation*. New York: John Wiley & Sons, 1986.

Brockmann describes the technical documentation process and convincingly argues for spending significant time during the planning phase. He also discusses the advantages and disadvantages of online versus paper documentation.

Grimm, Susan J. *How to Write Computer Manuals for Users*. Belmont, Calif.: Lifetime Learning Publications, 1982.

Although the subject is not broadly applicable, this book provides a very good discussion of the clear, direct writing it exemplifies.

Gunning, Robert. *The Technique of Clear Writing*. New York: McGraw-Hill Book Company, 1968.

Gunning describes the sins of complex, pretentious language and tells how not to sin. He also describes his Fog Index and other measures of readability. Gunning writes, "The price of good writing, as that of liberty, is eternal vigilance." What more can I say?

Horton, William K. *Designing and Writing Online Documentation: Help Files to Hypertext.* New York: John Wiley & Sons, 1990.

In great detail, Horton describes how to plan online documentation, from choosing a suitable structure to designing screens. Horton is the author of "The Wired Word," a column in *Technical Communication,* the Journal for the Society for Technical Communication, and is an outspoken authority on online documentation.

Strunk, William Jr., and E. B. White. *The Elements of Style.* New York: Macmillan Company, 1979.

An irreplaceable classic on the principles of good writing. Every writer must read this book at least once.

Zinsser, William. *On Writing Well.* New York: Harper Perennial, 1990.

Zinsser very informally touches on just about every aspect of writing nonfiction—simplicity and clutter, style, usage, audience, nonfiction as literature, what makes good nonfiction in various genres, and so on. He includes chapters on writing about science and technology and using a word processor.

INDEX

A

academic programs, 34–35, 37–38, 212–217 (table)
acronyms, defining, 70, 145
active listening, 86
active voice, 142–143
aerospace proposals, described, 32
agreement, between subject and verb, 158–159
alignment, defined, 139
American Medical Writer's Association (AMWA), address, 218
anthropomorphizing, avoiding, 148–149
appendixes, described, 121
aptitude, defined, 26
art
  in technical manuals, 129
  placing, 179
  spacing, 132
  tips on planning, 137–138
artists, communicating with, 135–137
assertiveness, 27–28
Association of Technical Writers and Editors, 17
Association for Business Communication, address, 218
Association of Teachers of Technical Writing, address, 219
audience, 92–104
  defining, 94–96
  in relation to visual elements, 129
  researching, 96–103

B

bibliography, described, 121
boldface, defined, 139
breaking in, 42–56
burnout, defined, 191

C

callouts

defined, 139
  fonts in, 133
capitalization, 161–163
  of list items, 162–163
  of titles, captions and headings, 162
career—See also career development
  books about, 226–228
  success, 200
career advancement, 200–201
career changers, 21
career development, 199–209
  consulting, 201–207
carpal tunnel syndrome, 195
  treatments for, 197
charts, as visual elements, 134
cold calls, in job hunting, 50–51
college degree, 34–38
  whether necessary, 20
colleges, 212–217 (table)
commas in a series, 160
communication—See also interviewing engineers
  in relation to stress, 192–193
  interpersonal, 24, 85–90
  visual, 24
  with artists, 135–137
  with editors, 167
  with engineers, 76–90
completeness, questions about, 99
components of technical books, 119–122
computer revolution, 18
computer tutorials, described, 33
conflict resolution, 176
consistency, 157, 161–163
  importance of, 157, 165
  in lists, 147, 162–163
consultant, defined, 201
consulting, 201–207